Classical Function Theory, Operator Dilation Theory, and Machine Computation on Multiply-Connected Domains

of the
American Mathematical Society

Number 892

Classical Function Theory,
Operator Dilation Theory,
and Machine Computation on
Multiply-Connected Domains

Jim Agler
John Harland
Benjamin J. Raphael

January 2008 • Volume 191 • Number 892 (second of 5 numbers) • ISSN 0065-9266

American Mathematical Society
Providence, Rhode Island

2000 *Mathematics Subject Classification.*
Primary 30–02, 46–02, 47–02.

Library of Congress Cataloging-in-Publication Data

Agler, Jim.
  Classical function theory, operator dilation theory, and machine computation on multiply-connected domains / Jim Agler, John Harland, Benjamin J. Raphael.
    p. cm. — (Memoirs of the American Mathematical Society, ISSN 0065-9266 ; no. 892)
  Includes bibliographical references.
  "Volume 191, number 892 (second of 5 numbers)."
  ISBN 978-0-8218-4046-7 (alk. paper)
  1. Geometric function theory. 2. Operator theory. 3. Dilation theory (Operator theory). 4. Functional analysis. 5. Analytic functions. I. Harland, John, 1959– II. Raphael, Benjamin J., 1974– III. Title.
QA331.A36 2007
515'.7—dc22   2007060555

Memoirs of the American Mathematical Society

This journal is devoted entirely to research in pure and applied mathematics.

Subscription information. The 2008 subscription begins with volume 191 and consists of six mailings, each containing one or more numbers. Subscription prices for 2008 are US$675 list, US$540 institutional member. A late charge of 10% of the subscription price will be imposed on orders received from nonmembers after January 1 of the subscription year. Subscribers outside the United States and India must pay a postage surcharge of US$38; subscribers in India must pay a postage surcharge of US$43. Expedited delivery to destinations in North America US$53; elsewhere US$130. Each number may be ordered separately; *please specify number* when ordering an individual number. For prices and titles of recently released numbers, see the New Publications sections of the *Notices of the American Mathematical Society*.

Back number information. For back issues see the *AMS Catalog of Publications*.

Subscriptions and orders should be addressed to the American Mathematical Society, P. O. Box 845904, Boston, MA 02284-5904, USA. *All orders must be accompanied by payment.* Other correspondence should be addressed to 201 Charles Street, Providence, RI 02904-2294, USA.

Copying and reprinting. Individual readers of this publication, and nonprofit libraries acting for them, are permitted to make fair use of the material, such as to copy a chapter for use in teaching or research. Permission is granted to quote brief passages from this publication in reviews, provided the customary acknowledgment of the source is given.

Republication, systematic copying, or multiple reproduction of any material in this publication is permitted only under license from the American Mathematical Society. Requests for such permission should be addressed to the Acquisitions Department, American Mathematical Society, 201 Charles Street, Providence, Rhode Island 02904-2294, USA. Requests can also be made by e-mail to `reprint-permission@ams.org`.

Memoirs of the American Mathematical Society is published bimonthly (each volume consisting usually of more than one number) by the American Mathematical Society at 201 Charles Street, Providence, RI 02904-2294, USA. Periodicals postage paid at Providence, RI. Postmaster: Send address changes to Memoirs, American Mathematical Society, 201 Charles Street, Providence, RI 02904-2294, USA.

© 2008 by the American Mathematical Society. All rights reserved.
Copyright of individual articles may revert to the public domain 28 years
after publication. Contact the AMS for copyright status of individual articles.
This publication is indexed in *Science Citation Index*®, *SciSearch*®, *Research Alert*®,
CompuMath Citation Index®, *Current Contents*®/*Physical, Chemical & Earth Sciences*.
Printed in the United States of America.

∞ The paper used in this book is acid-free and falls within the guidelines
established to ensure permanence and durability.
Visit the AMS home page at `http://www.ams.org/`

10 9 8 7 6 5 4 3 2 1   13 12 11 10 09 08

Contents

Preface	vii
Chapter 1. Generalizations of the Herglotz Representation Theorem, von Neumann's Inequality and the Sz.-Nagy Dilation Theorem to Multiply Connected Domains	1
1.1. Introduction	1
1.2. Preliminaries	7
1.3. The First Herglotz Representation	12
1.4. The Second Herglotz Representation	21
1.5. The Third Herglotz Representation	23
1.6. The Herglotz Representations and Operator Theory	28
1.7. An Application	33
Chapter 2. The Computational Generation of Counterexamples to the Rational Dilation Conjecture	61
2.1. Introduction	61
2.2. Mathematical Preliminaries	64
2.3. Analysis of the Dilation Condition for Nonsingularly Hyperextremal Grammians	72
2.4. Analysis of Dilation Extremal Grammians	82
2.5. Algorithms	89
2.6. A Computational Counterexample	100
2.7. Plausibility Arguments	105
Chapter 3. Arbitrary Precision Computations of the Poisson Kernel and Herglotz Kernels on Multiply-Connected Circle Domains	109
3.1. Introduction	109
3.2. Computation of the Functions	112
3.3. Results	119
Chapter 4. Schwartz Kernels on Multiply Connected Domains	127
Appendix A. Convergence Results	139
Appendix B. Example Inner Product Computation	155
Bibliography	157

Abstract

This work begins with the presentation of generalizations of the classical Herglotz Representation Theorem for holomorphic functions with positive real part on the unit disc to functions with positive real part defined on multiply-connected domains. The generalized Herglotz kernels that appear in these representation theorems are then exploited to evolve new conditions for spectral set and rational dilation conditions over multiply-connected domains. These conditions form the basis for the theoretical development of a computational procedure for probing a well-known unsolved problem in operator theory, the so called rational dilation conjecture. Arbitrary precision algorithms for computing the Herglotz kernels on circled domains are presented and analyzed. These algorithms permit an effective implementation of the computational procedure which results in a machine generated counterexample to the rational dilation conjecture.

2000 *Mathematics Subject Classification*. Primary: 30-02, 46-02, 47-02.

Key words and phrases. operator theory, Sz.-Nagy dilation theorem, spectral sets, rational dilation, Herglotz representations, Herglotz kernels, finitely connected domains, multiply connected domains, Schwartz kernels, Ahlfors functions, Blashke products.

This research was partially supported by NSF Grants DMS0100607 and DMS9801461. The authors are indebted to Scott McCullough, Bill Helton, and Mihai Putinar for numerous substantive conversations in connection with this work.

Preface

This memoir is divided into four chapters. In Chapter 1, we evolve three distinct generalizations of the Herglotz representation theorem from the unit disk to multiply-connected domains. These representation theorems are given in terms of certain Herglotz kernels associated to the domain. The representation theorems lead to generalizations of two classical operator-theoretic results on the unit disc to finitely-connected domains; namely, von Neumann's inequality and the Sz.-Nagy dilation theorem.

In Chapter 2, we employ the generalizations of von Neumann's inequality and the Sz.-Nagy dilation theory presented in Chapter 1 to evolve a computational approach to the *rational dilation conjecture*, which speculates that for each compact set $K \subseteq \mathbb{C}$, every bounded linear operator on Hilbert space that has K as a spectral set also has a rational dilation to ∂K. This approach yields specific 4×4 matrices, which we verify computationally are counterexamples for this conjecture in the case when K is the closure of a particular two holed domain.

A fundamental component of the computational procedure developed in Chapter 2 is the numerical computation of the Herglotz kernels introduced in Chapter 1. Because of the need for high precision, new algorithms were developed for the computation of these kernels (and the Poisson kernel as well) with arbitrary precision. Chapter 3 describes these algorithms in detail.

In Chapter 4 we relate the Herglotz kernels used in the first three chapters of the memoir to well known kernels that have appeared in various representation theorems in the literature.

The authors have structured the three chapters of the memoir so that each chapter can be read independently, according to the background and interests of the reader. Accordingly, Chapter 1 is written to be accessible both to function theorists unskilled in the nuances of operator theory, as well as operator theorists unaccustomed to the subtleties of function theory on multiply-connected domains. Similarly, Chapter 3 is written to be independently accessible to readers interested in the computational theory of multiply-connected domains, but who choose not to study the operator theoretic notions of Chapters 1 and 2 in detail. Finally, Chapter 2 is written with minimal reference to Chapters 1 and 3, and thus can in a large part be read independently by an expert in operator theory. As a consequence of this structure of the memoir, there is some overlap in the content of each chapter, most strikingly in separate introductions to each chapter. The expense of such overlap is justified by the desire for accessibility.

CHAPTER 1

Generalizations of the Herglotz Representation Theorem, von Neumann's Inequality and the Sz.-Nagy Dilation Theorem to Multiply Connected Domains

1.1. Introduction

Let K be a compact subset of the complex plane, $\mathbb{C}$, let $\mathcal{H}$ be a complex Hilbert space, and let $\mathcal{L}(\mathcal{H})$ denote the bounded linear transformations of $\mathcal{H}$ into $\mathcal{H}$. Following von Neumann [vN51] if $T \in \mathcal{L}(\mathcal{H})$ and $\sigma(T)$, the spectrum of T, is a subset of K, we say that *K is a spectral set for T* if

(1.1.1) $$\|f(T)\| \leq \max_{z \in K} |f(z)|$$

for all $f \in \mathcal{R}at(K)$, the algebra of rational functions with poles off K. Let $\mathcal{S}_K$ denote the operators that have K as a spectral set.

Now, if $T \in \mathcal{L}(\mathcal{H})$, how is it that T might have K as a spectral set? One particularly simple way was discovered by Sz.-Nagy in [SN53]. Let us agree to say that *T has a rational dilation to ∂K* if there exist a Hilbert space $\mathcal{K} \supseteq \mathcal{H}$ and a normal operator $N \in \mathcal{L}(\mathcal{K})$ such that $\sigma(N) \subseteq \partial K$ and

(1.1.2) $$f(T) = P_{\mathcal{H}} f(N) | \mathcal{H}$$

for all $f \in \mathcal{R}at(K)$. Let $\mathcal{D}_K$ denote the operators that have a rational dilation to ∂K. If N is a normal operator, then it follows immediately from the Spectral Theorem that

$$\|f(N)\| = \max_{z \in \sigma(T)} |f(z)|$$

for all $f \in \mathcal{R}at(\sigma(N))$. Hence, by the Maximum Principle, if $\sigma(N) \subset \partial K$ and $f \in \mathcal{R}at(K)$,

$$\|f(N)\| \leq \max_{z \in K} |f(z)|,$$

and we see immediately that if (1.1.2) holds then (1.1.1) holds.

Thus, if K is a compact subset of the plane, then $\mathcal{D}_K \subseteq \mathcal{S}_K$. Surprisingly, operator theory has been unable to find operators $T \in \mathcal{S}_K$ and that have K as spectral set for a reason different than in Sz.-Nagy's original example. Specifically, the following problem remains unresolved.

PROBLEM 1.1.3. *Give a concrete, "simple," model for an operator T for which by virtue of its presentation there is a particularly simple proof that $T \in \mathcal{S}_K$ and yet it is not obvious that $T \in \mathcal{D}_K$.*

The solution to Problem 1.1.3 is the Holy Grail of spectral set theory and we shall have little to say about it in this memoir. Rather, we shall focus on a closely related but more concrete and definite unsolved problem in the theory spectral sets, known as the dilation question.

PROBLEM 1.1.4. Is $\mathcal{D}_K = \mathcal{S}_K$?

Before continuing we remark that Problems 1.1.3 and 1.1.4 are independent. For example, it might be the case that it is shown that there exists a $T \in \mathcal{S}_K \setminus \mathcal{D}_K$ but that the proof is either nonconstructive[1] (so that there is no "simple" presentation of T) or by computer (so that there is no "simple" proof that $T \in \mathcal{S}_K$). Alternatively, perhaps it is shown that $\mathcal{S}_K = \mathcal{D}_K$, but the proof being quite complicated, Problem 1.1.3 would still represent a reasonable request. Finally, perhaps Problem 1.1.3 is solved satisfactorily in the following way. A concrete operator T is displayed, say via a function-theoretic model and the proof that $T \in \mathcal{S}_K$ is both brief and elementary. Yet the issue of whether $T \in \mathcal{D}_K$ reduces down to a delicate and unsolved problem in function theory that is thought to be quite hard. Of course it is also possible that one of the problems might be resolved in such a way that it resolves the other as well.

Operator theorists have discovered a great deal about Problem 1.1.4. By far the two most famous results in the theory are von Neumann's inequality [**vN51**] and the Sz.-Nagy dilation theorem [**SN53**].

THEOREM 1.1.5. *(von Neumann's Inequality) If $T \in \mathcal{L}(H)$ and $K = \mathbb{D}^-$, then $T \in \mathcal{S}_K$ if and only if $\|T\| \leq 1$.*

THEOREM 1.1.6. *(the Sz.-Nagy Dilation Theorem) If $T \in \mathcal{L}(H)$ and $K = \mathbb{D}^-$, then $T \in \mathcal{D}_K$ if and only if $\|T\| \leq 1$.*

As the reader can easily see by noticing the common condition in Theorems 1.1.5 and 1.1.6, $\|T\| \leq 1$, the above theorems immediately imply the solution to Problem 1.1.4 in the case when $K = \mathbb{D}^-$ (the answer is yes). Problem 1.1.4 has also been resolved in the affirmative in the cases when K has connected complement [**Sar65b**], when $R(K)$ is a Dirichlet algebra [**Leb63, Foi59, Ber68**], and when K is an annulus [**Agl85**], the bidisc [**And63**], or the symmetrized bidisc [**AY99, AY00**]. Other relevant and interesting results on dilation theory can be found, for instance, in [**Agl90, DP89, McC95, Mis84, Pau86, Pau87, Pau88, Put97a, Put97b, Put00**]. In this memoir we will not address Problem 1.1.4 directly. Rather, we shall give generalizations of Theorem 1.1.5 and Theorem 1.1.6 for more general K. These generalizations will, in turn, lead to an effective computational procedure for probing Problem 1.1.4 for a large class of K.

Specifically, we shall assume $K = \Omega^-$, where Ω is a bounded connected open set in the complex plane whose boundary consists of $m+1$ disjoint simple analytic closed curves $\partial_0, \ldots, \partial_m$. Define an $m+1$-dimensional torus, T_Ω, by setting

$$T_\Omega = \partial_0 \times \partial_1 \times \cdots \times \partial_m = \{\alpha = (\alpha_r) : \alpha_r \in \partial_r \text{ for } r = 0, \ldots, m\}.$$

To extend Theorem 1.1.5 to Ω we define for each $\alpha \in T_\Omega$ a holomorphic function on Ω, $\Gamma_\alpha(z)$, in the following way. Fix $z_0 \in \Omega$, and let σ be the harmonic measure on $\partial \Omega$ for the point z_0. There exists a Poisson kernel $P_z(\lambda)$ such that given any continuous function $u : \partial \Omega \to \mathbb{R}$, the harmonic function $\hat{u} : \Omega \to \mathbb{R}$ defined by

$$(1.1.7) \qquad \hat{u}(z) = \int_{\partial \Omega} P_z(\lambda) u(\lambda) \, d\sigma(\lambda)$$

---

[1] While the present monograph was under review, Dritschel and McCullough [**DM05**] provided such a nonconstructive proof.

solves the Dirichlet problem on Ω with boundary data u. It turns out that for each $\alpha \in T_\Omega$ there exists an $m+1$-tuple of positive weights $(\omega_0(\alpha), \ldots, \omega_m(\alpha))$ such that

$$\sum_{r=0}^{m} w_r(\alpha) = 1 \tag{1.1.8}$$

and such that the harmonic function $u_\alpha : \Omega \to \mathbb{R}$ defined by

$$u_\alpha(z) = \sum_{r=0}^{m} w_r(\alpha) P_z(\alpha_r)$$

has a single-valued harmonic conjugate on Ω. Let $\Gamma_\alpha(z)$ denote the unique holomorphic function on Ω with

$$Re\, \Gamma_\alpha = u_\alpha \tag{1.1.9}$$

and

$$Im\, \Gamma_\alpha(z_0) = 0. \tag{1.1.10}$$

Note that since σ is harmonic measure for the point z_0, (1.1.7) implies that $P_{z_0}(\lambda) = 1$ for all $\lambda \in \partial\Omega$. Hence from (1.1.8), (1.1.9), and (1.1.10) it follows that

$$\Gamma_\alpha(z_0) = 1 \quad \text{for all } \alpha \in T_\Omega. \tag{1.1.11}$$

Furthermore, since $P_z(\lambda) > 0$ for all $z \in \Omega$ and $\lambda \in \partial\Omega$, and since the weights $w_0(\alpha), \ldots, w_m(\alpha)$ are positive,

$$Re\, \Gamma_\alpha(z) > 0 \quad \text{for all } z \in \Omega \text{ and all } \alpha \in T_\Omega. \tag{1.1.12}$$

We remark that in the main body of the memoir we give a slightly more general presentation of the functions Γ_α. In particular, we replace σ with a more general measure and we replace the condition (1.1.10) with a more general linear condition on $Im\, \Gamma_\alpha$.

The functions Γ_α, as defined above, have been studied by F. Forelli [**For79**], H. Grunsky [**Gru78**], M. Heins [**Hei85**], and S. Fisher and A. Khavinson [**FK99**], and have a number of characterizations. From the point of view of functional analysis, they are the extreme points of the convex set C of holomorphic functions on Ω defined by

$$C = \{f : \Omega \to \mathbb{C} : f \text{ is holomorphic}, Re\, f \geq 0, \text{ and } f(z_0) = 1\}. \tag{1.1.13}$$

From the point of view of classical function theory, they are the $m+1$-valent functions f from Ω onto the right half plane, normalized so that $f(z_0) = 1$. Alternatively, we can compose Γ_α with he conformal mapping $\frac{w-1}{w+1}$ from the right half plane onto the unit disk to obtain a function $\phi_\alpha : \Omega \to \mathbb{D}$ defined by

$$\phi_\alpha(z) = \frac{\Gamma_\alpha(z) - 1}{\Gamma_\alpha(z) + 1}.$$

The functions ϕ_α, $\alpha \in T_\Omega$ are precisely the single-valued Blashke products of degree $m+1$ (or, equivalently, of minimal degree) on Ω which map z_0 to 0 (see, for example, [**Fis83**] for an exposition of Blashke products).

We now give our generalization of Theorem 1.1.5 to Ω (this appears as Theorem 5.4 in the body of the memoir).

THEOREM 1.1.14. *If $T \in \mathcal{L}(\mathcal{H})$, $\sigma(T) \subseteq \Omega$, and $K = \Omega^-$ then $T \in \mathcal{S}_K$ if and only if*
$$\operatorname{Re} \Gamma_\alpha(T) \geq 0 \qquad \text{for all } \alpha \in T_\Omega.$$

To see that Theorem 1.1.14 is indeed a generalization of Theorem 1.1.5, notice that when $\Omega = \mathbb{D}$, $T_\Omega = \partial_0 = \partial \mathbb{D}$, and if we let $z_0 = 0$ then
$$\Gamma_\alpha(z) = \frac{\alpha_0 + z}{\alpha_0 - z}.$$
Assuming $\sigma(T) \subset \mathbb{D}$, we have
$$\operatorname{Re} \Gamma_\alpha(T)$$
$$= \frac{1}{2}\left(\frac{\alpha_0 + T}{\alpha_0 - T} + \left(\frac{\alpha_0 + T}{\alpha_0 - T}\right)^*\right)$$
$$= \frac{1}{2}\left(\frac{1}{\alpha_0 - T}\right)^* \left((\alpha_0 - T)^*(\alpha_0 + T) + (\alpha_0 + T)^*(\alpha_0 - T)\right)\left(\frac{1}{\alpha_0 - T}\right)$$
$$= \left(\frac{1}{\alpha_0 - T}\right)^*(1 - T^*T)\left(\frac{1}{\alpha_0 - T}\right)$$
and thus $\operatorname{Re} \Gamma_\alpha(T) \geq 0$ for all $\alpha \in T_\Omega$ if and only if $1 - T^*T \geq 0$, which is equivalent to the assertion $\|T\| \leq 1$. Thus, Theorem 1.1.14 implies 1.1.5 in the special case where $\sigma(T) \subseteq \mathbb{D}$. To remove this restriction and obtain the full strength of Theorem 1.1.5, one can either employ the simple trick of replacing T by rT, where $0 < r < 1$, and then letting $r \to 1$, or interpret $\operatorname{Re} \Gamma_\alpha(T)$ as an unbounded operator via a Cayley transform argument.

To extend Theorem 1.1.6 to Ω we define, for each $\lambda \in \partial \Omega$, a holomorphic function on Ω, $H_\lambda(z)$, in the following way. Let $\{\phi_1, \ldots, \phi_m\}$ be an orthonormal basis for the orthogonal complement of $\{\operatorname{Re} f : f \in \mathcal{R}at(\Omega^-)\}$ in $L^2_\mathbb{R}(\sigma)$, the set of real-valued functions on $\partial \Omega$ which are square integrable with respect to σ. It turns out that for each $\lambda \in \partial \Omega$, the harmonic function $v_\lambda : \Omega \to \mathbb{R}$ defined by

$$(1.1.15) \qquad v_\lambda(z) = P_z(\lambda) - \sum_{r=1}^m \phi_r(\lambda)\hat{\phi}_r(z)$$

has a single-valued harmonic conjugate on Ω. Let $H_\lambda(z)$ denote the unique holomorphic function on Ω with

$$(1.1.16) \qquad \operatorname{Re} H_\lambda = v_\lambda$$

and
$$\operatorname{Im} H_\lambda(z_0) = 0.$$
We now give our generalization of Theorem 1.1.6 (this appears as Theorem 1.6.18 in the body of the memoir).

THEOREM 1.1.17. *If $T \in \mathcal{L}(H)$, $\sigma(T) \subseteq \Omega$, and $K = \Omega^-$ then $T \in \mathcal{D}_K$ if and only if there exist self-adjoint $\Phi_1, \ldots, \Phi_m \in \mathcal{L}(H)$ such that*

$$(1.1.18) \qquad \operatorname{Re} H_\lambda(T) + \sum_{r=1}^m \phi_r(\lambda)\Phi_r \geq 0 \qquad \text{for all } \lambda \in \partial \Omega.$$

To see that Theorem 1.1.17 is a generalization of Theorem 1.1.6, observe that if $\Omega = \mathbb{D}$ and $z_0 = 0$, then $m = 0$ and
$$H_\lambda(\Omega) = \frac{\lambda + z}{\lambda - z}.$$

The condition (1.1.18) hence reduces to

$$Re \frac{\lambda+T}{\lambda-T} \geq 0 \qquad \text{for all } \lambda \in \partial\mathbb{D},$$

which has already been shown to be equivalent to $\|T\| \leq 1$ when $\sigma(T) \subseteq \mathbb{D}$. The restriction $\sigma(T) \subseteq \mathbb{D}$ can be removed as before.

In summary, when $\Omega = \mathbb{D}$, the functions $\Gamma_\alpha(z)$ and $H_\lambda(z)$ appearing in Theorems 1.1.14 and 1.1.17 reduce to the same function, namely the classical Herglotz kernel $\frac{\lambda+z}{\lambda-z}$. $\Gamma_\alpha(z)$ and $H_\lambda(z)$ are thus two different generalizations of the classical Herglotz kernel on the disc to finitely connected domains Ω. Accordingly, we shall refer to $\Gamma_\alpha(z)$ as the *Herglotz kernel of the first kind* and to $H_\lambda(z)$ as the *Herglotz kernel of the second kind*.

Now, there is more than just a Herglotz kernel on the disc; there is also the Herglotz theorem.

THEOREM 1.1.19. *(the Herglotz Representation Theorem) f is a holomorphic function on $\mathbb{D}$ with $Re\, f \geq 0$ and $f(0) = 1$ if and only if there exists a probability measure μ on $\partial\mathbb{D}$ such that*

$$(1.1.20) \qquad f(z) = \int_{\partial\mathbb{D}} \frac{\lambda+z}{\lambda-z} d\mu(\lambda) \qquad \text{for all } z \in \mathbb{D}.$$

Moreover if f is a holomorphic function on $\mathbb{D}$ with $Re\, f \geq 0$ and $f(0) = 1$, then there exists a unique μ such that (1.1.20) holds.

It has long been known that there is an intimate connection between the Herglotz Representation Theorem and the Sz.-Nagy Dilation Theorem (see e.g. [**Fil70**]). Here we content ourselves with pointing out that there is a particularly transparent proof of von Neumann's Inequality using the Herglotz Representation Theorem. Accordingly, assume that $T \in \mathcal{L}(\mathcal{H})$ and $\|T\| \leq 1$. By a Cayley Transform argument von Neumann's Inequality will follow if we can show that

$$Re\, f(rT) \geq 0$$

whenever $Re\, f \geq 0$ on $\mathbb{D}$, $f(0) = 1$, and $r < 1$. But if $Re\, f \geq 0$ on $\mathbb{D}$ and $f(0) = 1$, Theorem 1.1.19 implies that there exists a positive measure μ such that

$$f(z) = \int \frac{e^{i\theta}+z}{e^{i\theta}-z} d\mu(e^{i\theta})$$

and also since $\|T\| \leq 1$ and $r < 1$,

$$Re \frac{e^{i\theta}+rT}{e^{i\theta}-rT} \leq 0$$

for all θ. Hence

$$Re\, f(rT) = Re \int \frac{e^{i\theta}+rT}{e^{i\theta}-rT} d\mu(e^{i\theta})$$
$$= \int Re \frac{e^{i\theta}+rT}{e^{i\theta}-rT} d\mu(e^{i\theta})$$
$$\geq 0.$$

This proves von Neumann's Inequality.

It was the desire to generalize this argument to Ω that led the authors to attempt to generalize the Herglotz representation theorem to Ω. However, we found that when Ω has nontrivial genus, there is no single, canonical way to evolve such a generalization. Indeed, in this memoir we shall prove three distinct representation theorems for certain classes of holomorphic functions on Ω, all of which reduce to Theorem 1.1.19 when $\Omega = \mathbb{D}$. The first such result, which is used to prove Theorem 1.1.14, is based on an extreme point analysis of the set C of holomorphic functions defined by (1.1.13) (this theorem appears as Theorem 1.3.27 in Section 1.3).

THEOREM 1.1.21. *(the first Herglotz representation) $f \in C$ if and only if there exists a probability measure μ on T_Ω such that*

$$(1.1.22) \qquad f(z) = \int_{T_\Omega} \Gamma_\alpha(z) \, d\mu(\alpha) \qquad \text{for all } z \in \Omega.$$

The second generalization of the Herglotz theorem, which will be used to prove Theorem 1.1.17, again provides a representation for elements of C, but is based on the idea of expressing the interior values of a holomorphic function on Ω in terms of the boundary values of its real part (this result appears as Theorem 1.4.5 in Section 1.4).

THEOREM 1.1.23. *(the second Herglotz representation) $f \in C$ if and only if there exists a probability measure μ on $\partial\Omega$ such that*

$$(1.1.24) \qquad \int_{\partial\Omega} \hat{\phi}_r \, d\mu = 0 \qquad \text{for all } r = 1, \ldots, m$$

and

$$(1.1.25) \qquad f(z) = \int_{\partial\Omega} H_\lambda(z) \, d\mu(\lambda) \qquad \text{for all } z \in \Omega.$$

Furthermore, if $f \in C$ then there exist a unique μ such that (1.1.24) and (1.1.25) hold.

Representation theorems for functions on multiply connected domains which are similar to the above result have appeared in the literature—see, for example, [**CW67**], [**FK99**], and [**Kha84**]. The relationship between Theorem 1.1.23 and such related results is given in Chapter 4.

Note that in the classical Herglotz theorem, Theorem 1.1.19, the measure μ is unique and every probability measure is admissible. In contrast, Theorems 1.1.21 and 1.1.23 each carry a proviso: in Theorem 1.1.21, every probability measure is admissible but is not necessarily unique, and in Theorem 1.1.23 the measure is unique but, since (1.1.24) is imposed, not every probability measure is admissible. The following result, our third generalization of the Herglotz theorem, removes these shortcomings, but it does so at the expense of representing a larger class of functions than C. Specifically, we define a set C_1 of holomorphic functions on Ω by setting

$(1.1.26)\quad C_1 = \{f : \Omega \to \mathbb{C} : f \text{ is holomorphic, } f(z_0) = 1, \text{ and there exist}$

$$c_1, \ldots, c_m \in \mathbb{R} \text{ such that } \operatorname{Re} f(z) + \sum_{r=1}^{m} c_r \hat{\phi}_r(z) > 0 \text{ for all } z \in \Omega.\}$$

In Section 1.5 we shall show that C_1 is a canonical object from the point of view of functional analysis and in addition show that the functions H_λ are the extreme points of C_1. This will lead to the following result (which appears as Theorem 1.5.14 in the body of this memoir).

THEOREM 1.1.27. *(the third Herglotz representation)* $f \in C_1$ *if and only if there exists a probability measure μ on $\partial\Omega$ such that*

$$(1.1.28) \qquad f(z) = \int_{\partial\Omega} H_\lambda(z)\, d\mu(\lambda) \qquad \text{for all } z \in \Omega.$$

Furthermore, f and μ determine each other uniquely via (1.1.28).

This chapter is divided into six sections. In Section 1.2 we introduce notation and collect together various well known facts from function theory on multiply connected domains and from functional analysis that will be used throughout the memoir. Section 1.3 introduces the Herglotz kernel of the first kind, characterizes the kernel canonically from a number of points of view, and proves the first Herglotz representation. Section 1.4 introduces the Herglotz kernel of the second kind and proves the second Herglotz representation. Section 1.5 characterizes the Herglotz kernel of the second kind canonically and establishes the third Herglotz representation. Section 1.6 applies the first and second Herglotz representations to prove Theorems 1.1.14 and 1.1.17, respectively. Section 1.7 presents a concrete application of the results of Sections 1.3–1.6, namely a computational method for probing Problem 1.1.4 when $K = \Omega^-$, the closure of a multiply connected domain with analytic boundary. This method is extended in Chapter 2, where a counterexample to the dilation question is presented.

1.2. Preliminaries

In this section we shall present the basic concepts from the theory of analytic functions on multiply connected domains in the plane that we shall require in the subsequent sections of the memoir. For the benefit of readers unskilled in the nuances of function theory on multiply connected domains we have written the section in semi-expository style. Experts need only scan the section for the notation which will be adhered to throughout the remainder of the memoir.

For X a topological space and $S \subseteq X$ we let S^0 denote the interior of S, S^- denote the closure of S and ∂S denote the boundary of S. If, in addition, X is compact and Hausdorff, we let $C(X)$ denote the Banach space of continuous complex-valued functions on X and $M(X)$ the Banach space of finite regular complex Borel measures on X. Similarly, $C_\mathbb{R}(X)$ will denote the continuous real-valued functions on X and $M_\mathbb{R}(X)$ will denote the space of finite regular real Borel measures. We let $Prob(X)$ denote the probability measures on X. For G an open set in the plane let $Hol(G)$ denote the space of holomorphic functions on G and let $Harm(G)$ denote the space of harmonic complex-valued functions on G. Thus, in general $Hol(G) + \overline{Hol(G)} \subseteq Harm(G)$, and the containment is an equality if and only if G is simply connected. Also, if we let $Harm_\mathbb{R}(G)$ denote the real valued harmonic functions on G, then $\operatorname{Re} f \in Harm_\mathbb{R}(G)$ whenever $f \in Hol(G)$.

Let Ω be a bounded connected open set in the plane whose boundary consists of $m+1$ disjoint simple analytic closed curves $\partial_0, \partial_1, \ldots, \partial_m$. It is well known that if $u \in C_\mathbb{R}(\partial\Omega)$, then there exists a unique function $u^\wedge \in C_\mathbb{R}(\Omega^-)$ such that $u^\wedge \mid \Omega \subseteq Harm_\mathbb{R}(\Omega)$ and

$$(1.2.1) \qquad u^\wedge \mid \partial\Omega = u.$$

We say that $u^\wedge$ is the solution to the Dirichlet problem with boundary data u. We wish to extend this process to the case where u is a measure. Accordingly, we must adopt a convention by which elements $u \in C_\mathbb{R}(\partial\Omega)$ may be viewed as elements of $M_\mathbb{R}(\partial\Omega)$.

Specifically, fix $\sigma \in M_\mathbb{R}(\partial\Omega)$. For technical simplicity we assume $d\sigma = w(\lambda)\,ds$ where ds is arclength measure on $\partial\Omega$ and w is a strictly positive continuous function on $\partial\Omega$ such that $\int_{\partial\Omega} w\,ds = 1$. There exists a function $P_z^\sigma(\lambda)$, *the Poisson kernel for Ω with respect to σ*, defined for all $z \in \Omega$ and $\lambda \in \partial\Omega$, with $P_z^\sigma \in C_\mathbb{R}(\partial\Omega)$ for each $z \in \Omega$, $z \mapsto P_z^\sigma(\lambda)$ harmonic on Ω for each $\lambda \in \partial\Omega$, and such that

$$(1.2.2) \qquad u^\wedge(z) = \int_{\partial\Omega} P_z^\sigma(\lambda) u(\lambda)\,d\sigma(\lambda)$$

whenever $u \in C_\mathbb{R}(\partial\Omega)$ and $z \in \Omega$.

Thus, if we identify the *function* $u \in C_\mathbb{R}(\partial\Omega)$ with the *measure* $u d\sigma \in M_\mathbb{R}(\partial\Omega)$, then a consistent way to define $\mu^\wedge \in Harm_\mathbb{R}(\Omega)$ for $\mu \in M_\mathbb{R}(\partial\Omega)$ would be given by the following formula.

$$(1.2.3) \qquad \mu^\wedge(z) = \int_{\partial\Omega} P_z^\sigma(\lambda)\,d\mu.$$

Thus, if $\mu = u d\sigma$, then $\mu^\wedge = u^\wedge$. Also, note that since P_z^σ is continuous for each $z \in \Omega$, (1.2.3) gives the *only* extension of (1.2.2) to measures (with the convention that u is identified with $u d\sigma$) with the property that $\mu \to \mu^\wedge(z)$ is $wk*$ continuous for each $z \in \Omega$.

We summarize three elementary facts about the map $\mu \mapsto \mu^\wedge$ which will be used repeatedly in the sequel. First, observe that if $\lambda \in \partial\Omega$ and δ_λ denotes the unit point mass concentrated at λ (i.e. $\delta_\lambda(\{\lambda\}) = 1$ and $\delta_\lambda(\Delta) = 0$ if $\Delta \subseteq \partial\Omega\setminus\{\lambda\}$), then

$$(1.2.4) \qquad \delta_\lambda^\wedge(z) = P_z^\sigma(\lambda).$$

Second,

$$(1.2.5) \qquad \text{if } \{\mu_\ell\} \subseteq M_\mathbb{R}(\Omega) \text{ is a sequence and } \mu_\ell \to \mu \in M_\mathbb{R}(\Omega)$$
$$\text{in the } wk* \text{ topology, then } \mu_\ell^\wedge \text{ converges to}$$
$$\mu^\wedge \text{ uniformly on compact subsets of } \Omega.$$

Finally, we point out that there is a very real sense in which (1.2.1) holds not just for $u \in C_\mathbb{R}(\partial\Omega)$ but for measures as well. Choose a sequence of continuous 1-1 maps $\tau_\ell : \partial\Omega \to \Omega$ with the property that

$$\lim_{\ell \to \infty} \max_{\lambda \in \partial\Omega} |\tau_\ell(\lambda) - \lambda| = 0.$$

Thus, if u is a continuous real valued function on Ω, one can consider the function $u^\wedge \circ \tau_\ell \in C_\mathbb{R}(\partial\Omega)$ and, as before, identify it with the measure $(u^\wedge \circ \tau_\ell)d\sigma \in M_\mathbb{R}(\partial\Omega)$. Since $u^\wedge \in C_\mathbb{R}(\Omega^-)$ whenever $u \in C_\mathbb{R}(\partial\Omega)$ it is clear that (1.2.1) is equivalent to the assertion that $u^\wedge \circ \tau_\ell \to u$ in $C_\mathbb{R}(\partial\Omega)$. When $\mu \in M_\mathbb{R}(\partial\Omega)$, we have that

$$(1.2.6) \qquad (\mu^\wedge \circ \tau_\ell)d\sigma \to \mu \text{ in the } wk* \text{ topology.}$$

Now recall that a harmonic function $u \in Harm_\mathbb{R}(\Omega)$ is said to have a harmonic conjugate on Ω if there exists a function v on Ω such that $u + iv \in Hol(\Omega)$. Furthermore if a harmonic conjugate exists, then it is unique up to a real constant. In particular, if $z_0 \in \Omega$ and u is a continuous real valued function on Ω that has a harmonic conjugate on Ω, then there exists a unique harmonic conjugate v such that $v(z_0) = 0$. Also, if $u \in C_\mathbb{R}(\partial\Omega)$ we say *u has a harmonic conjugate on* Ω if $u^\wedge$ has a harmonic conjugate on Ω, a slight and harmless abuse of language since u and $u^\wedge$ determine each other uniquely. Likewise, if $\mu \in M_\mathbb{R}(\partial\Omega)$, let us agree to say μ *has a harmonic conjugate on* Ω if $\mu^\wedge$ has a harmonic conjugate on Ω. In light of the observation immediately following (1.2.3) it should be clear that if $u \in C_\mathbb{R}(\partial\Omega)$, then $ud\sigma$ has a harmonic conjugate if and only if u has a harmonic conjugate so that our language is consistent.

In this memoir if X is a space of real valued objects on Ω or $\partial\Omega$, we shall adopt the convention of using a superscript h to denote the subspace of X consisting of all elements of X that have a harmonic conjugate on Ω. Thus, $Harm_\mathbb{R}^h(\Omega)$ denotes the space of real valued harmonic functions on Ω that have a harmonic conjugate on Ω, $C_\mathbb{R}^h(\partial\Omega)$ denotes the space of continuous real valued functions on $\partial\Omega$ that have a harmonic conjugate on Ω, $M_\mathbb{R}^h(\partial\Omega)$ denotes the space of real finite regular Borel measures on $\partial\Omega$ that have a harmonic conjugate, and $Prob^h(\partial\Omega) = M_\mathbb{R}^h(\partial\Omega) \cap Prob(\partial\Omega)$.

Now, a basic fact is that $Harm_\mathbb{R}^h(\Omega)$ has codimension exactly m in $Harm_\mathbb{R}(\Omega)$. To nail down how $Harm_\mathbb{R}^h(\Omega)$ sits inside $Harm_\mathbb{R}(\Omega)$, let $L_\mathbb{R}^2(\sigma)$ denote the real Hilbert space of measurable square integrable real valued functions on $\partial\Omega$ with inner product given by

$$\langle u, v \rangle = \int_{\partial\Omega} u(\lambda)v(\lambda)d\sigma(\lambda)$$

and, following our convention, let $L_\mathbb{R}^{2,h}(\sigma)$ denote the subspace of elements $u \in L_\mathbb{R}^2(\sigma)$ such that $ud\sigma$ has a harmonic conjugate on Ω. It turns out that $(L_\mathbb{R}^{2,h}(\sigma))^\perp$, the orthogonal complement of $L_\mathbb{R}^{2,h}(\sigma)$ in $L_\mathbb{R}^2(\sigma)$ has dimension m and consists entirely of continuous functions. Accordingly, we can fix an orthonormal basis for $L_\mathbb{R}^{2,h}(d\sigma)^\perp$ consisting of the m continuous functions $\phi_1, \phi_2, \ldots, \phi_m$.

We now recall some elementary results from functional analysis. If X is a real Banach space and $E \subseteq X$, let us agree to say that E *is a wedge* if $x + y \in X$ and $tx \in X$ whenever $x, y \in X$ and $t \geq 0$. If $E \subseteq X$ let us agree to let $E^\perp$ denote the collection of all $x^* \in X^*$, the dual of X, such that $x^*(e) \geq 0$ for all $e \in E$. Similarly, if $E \subseteq X^*$ let $^\perp E$ denote the collection of all $x \in X$ such that $e(x) \geq 0$ for all $e \in E$. Evidently, if $E \subseteq X$, then $E^\perp$ is wedge that is closed in the $wk*$ topology on X^* and if $E \subseteq X^*$, then $^\perp E$ is wedge that is closed in the norm topology on X. Simple consequences of the Hahn–Banach Theorem are that $^\perp(E^\perp)$ is the closed wedge generated by E (i.e. minimal in the set of closed wedges containing E) whenever $E \subseteq X$ and that $(^\perp E)^\perp$ is the $wk*$ closed wedge generated by E whenever $E \subseteq X^*$. In particular,

(1.2.7) $\qquad ^\perp(E^\perp) = E \qquad$ if E is a closed wedge in X,

and

(1.2.8) $\qquad (^\perp E)^\perp = E \qquad$ if E is $wk*$ closed wedge in X^*,

facts that we shall use in the sequel.

In terms of the $E^\perp$ notation it is now easy to describe how the spaces $M_\mathbb{R}^h(\partial\Omega)$ and $C_\mathbb{R}^h(\partial\Omega)$ are related to the orthonormal basis $\phi_1, \ldots, \phi_m$ for $L_\mathbb{R}^{2,h}(d\sigma)^\perp$. Recall that $C_\mathbb{R}(\partial\Omega)$ is a real Banach space and that $C_\mathbb{R}(\partial\Omega)^*$ can be identified with

$M_\mathbb{R}(\partial\Omega)$ via the pairing $(u,\mu) \mapsto \int u d\mu$.

(1.2.9) $$M_\mathbb{R}^h(\partial\Omega) = \{\phi_1,\ldots,\phi_m\}^\perp.$$

(1.2.10) $$C_\mathbb{R}^h(\partial\Omega)^\perp = \text{span}\{\phi_1 d\sigma,\ldots,\phi_m d\sigma\}.$$

The pathology just analyzed for nonsimply connected domains, that not all harmonic functions necessarily have a harmonic conjugate, is relatively tame by comparison to the problem of describing the boundary values of harmonic functions. Thus, the limit described in (1.2.6) fails to exist in general when $\mu^\wedge$ is replaced by an arbitrary $u \in Harm_\mathbb{R}(\Omega)$. Equivalently, not every $u + Harm_\mathbb{R}(\Omega)$ has the form $u = \mu^\wedge$ for some $\mu \in M_\mathbb{R}(\Omega)$. Indeed, the harmonic functions with the above properties form set of the first category in $Harm_\mathbb{R}(\Omega)$, a complete metric space in the topology of uniform convergence on compact subsets. All the more remarkable then is the following property of the $Harm_\mathbb{R}(\Omega)$.

If $u \in Harm_\mathbb{R}(\Omega)$ and $u \geq 0$; then there exists

(1.2.11) $\mu \in M_\mathbb{R}(\partial\Omega)$ such that $u \circ \tau_\ell d\sigma \to \mu$ in the $wk*$ topology.

Furthermore, $\mu \geq 0$ and $u = \mu^\wedge$.

We can combine (1.2.5), (1.2.6) and (1.2.11) to give an elegant parameterization of the normalized positive harmonic functions on Ω. Recall that $Harm_\mathbb{R}(\Omega)$ is a locally convex topological vector space when equipped with the topology of uniform convergence on compact subsets of Ω. Also, if $u \in Harm_\mathbb{R}(\Omega)$ and $u \geq 0$, then it follows from (1.2.11) that $\sigma(u) \in \mathbb{R}$ can be defined by

$$\sigma(u) = \lim_{\ell \to \infty} \int u(\tau_\ell(\lambda)) d\sigma(\lambda).$$

LEMMA 1.2.12. *The map $\mu \mapsto \mu^\wedge$ is a continuous 1-1 linear mapping from $(M_\mathbb{R}(\partial\Omega), wk*)$ into $Harm_\mathbb{R}(\Omega)$. The map $\mu \mapsto \mu^\wedge$ is an affine homeomorphism from $(Prob(\partial\Omega), wk*)$ onto $\{u \in Harm(\Omega) : u \geq 0 \text{ and } \sigma(u) = 1\}$.*

PROOF. That $\mu \mapsto \mu^\wedge$ is a 1-1 continuous linear map from $M_\mathbb{R}(\partial\Omega)$ into $Harm(\Omega)$ follows from (1.2.5) and (1.2.6). That $\mu \mapsto \mu^\wedge$ is a bijection from $Prob(\partial\Omega)$ onto $\{u \in Harm(\Omega) : u \geq 0 \text{ and } \sigma(u) = 1\}$ follows from (1.2.11). Finally, that $\mu \mapsto \mu^\wedge$ is a homeomorphism of $(Prob(\partial\Omega), wk*)$ follows from the fact that $(Prob(\partial\Omega), wk*)$ is compact. $\square$

Before continuing we make a subtle but elementary remark concerning the map $\mu \mapsto \mu^\wedge$ of Lemma 1.2.12: it is clear from (1.2.2) and (1.2.3) that this map in fact depends on σ. Accordingly, we introduce the following notation.

DEFINITION 1.2.13. For σ a normalizing measure as in the discussion leading up to (1.2.2), we define $I_\sigma : M_\mathbb{R}(\partial\Omega) \to Harm(\Omega)$ by the formula,

$$I_\sigma(\mu) = \mu^\wedge.$$

If σ_1 and σ_2 are two normalizing measures then how are I_{σ_1} and I_{σ_2} related? From (1.2.2) and the fact that $\{P_z^{ds} : z \in \Omega\}$ has dense span in $C_\mathbb{R}(\partial\Omega)$ it is easy to see that the Poisson kernels attached to σ_1 and σ_2 are related by the following formula.

(1.2.14) $$P_z^{\sigma_2}(\lambda) = \frac{d\sigma_1}{d\sigma_2}(\lambda) P_z^{\sigma_1}(\lambda) \quad \text{if } z \in \Omega \text{ and } \lambda \in \partial\Omega.$$

Thus, from (1.2.3),

(1.2.15) $$I_{\sigma_2}(\mu) = I_{\sigma_1}\left(\frac{d\sigma_1}{d\sigma_2}\mu\right) \quad \text{if } \mu \in M_{\mathbb{R}}(\partial\Omega).$$

Because of the trivial nature of the transformation formulas (1.2.14) and (1.2.15), function theorists commonly make a specific choice for the normalizing measure σ. The choices seen most frequently are $\sigma = ds$, arclength (possibly normalized), which results in the self-adjointness of the Poisson kernel, and $\sigma = dm_{z_0}$, harmonic measure for the point z_0, which has nice transformation properties under conformal mappings, covering mappings, and Ahlfor's mappings. There is, however, nothing "canonical" about these choices. Rather, they simply serve to make certain formulas take on a simpler appearance. In the authors' particular approach to the function theory on Ω via machine computation the simplest choice of σ turns out to be to let $\sigma = \Sigma\sigma_r$ where σ_r is normalized arclength measure on ∂_r, a choice which is general is different than the two just alluded to. For these reasons, that there is no way to make a "canonical" choice of σ and that one does not in general know in advance of an application what the "simple" choice of σ may be, the authors have elected to carry the slight generality of the weight σ through their theoretical discussion.

We now wish to introduce notation that will allow us to clearly parameterize certain classes of holomorphic functions in terms of their real parts. Following the approach elucidated in the previous paragraph we shall do this in a way that is trivially more general than the usual approach. Usually, one fixes a point $z_0 \in \Omega$, notes that each $u \in Harm^h(\Omega)$ has a unique harmonic conjugate v such that $v(z_0) = 0$, and then introduces the linear operator $H : Harm^h(\Omega) \to Harm(\Omega)$ defined by requiring that $u + iH(u)$ be holomorphic on Ω and $H(u)(z_0) = 0$. H is referred to as the *harmonic conjugation operator*. It is easy to see that H is linear, 1-1, and that $ran\, H = \{v \in Harm^h(\Omega) : v(z_0) = 0\}$. In addition, it is easy to see from the Cauchy–Riemann equations that H is a continuous mapping when $Harm^h(\Omega)$ is equipped with the topology of uniform convergence on compact subsets of Ω.

We consider the following more general strategy for picking a harmonic conjugate. By an *affine mapping* of a vector space V, we shall mean a mapping $\tau : V \to W$ where W is a vector space and where τ has the special form

(1.2.16) $$\tau(v) = L(v) + w_0, \quad v \in V$$

where $L : V \to W$ is linear and $w_0 \in W$. If V is vector space over $\mathbb{F}$ and $W = \mathbb{F}$, then we say that τ is an *affine functional*. Now, fix an affine functional on $Harm_{\mathbb{R}}(\Omega)$, say $\tau = L+c$ with L linear and $c \in \mathbb{R}$. We define the *harmonic conjugation operator with respect to* τ, $H_\tau : Harm^h(\Omega) \to Harm(\Omega)$ by requiring

(1.2.17) $$u + iH_\tau(u) \text{ is holomorphic on } \Omega \text{ and } \tau(H_\tau(u)) = 0.$$

for all $u \in Harm^h(\Omega)$. To analyze H_τ we note that since $u+iH_\tau(u)$ is holomorphic on Ω, for each $u \in Harm^h(\Omega)$, there must exist $s \in \mathbb{R}$ such that

(1.2.18) $$H_\tau(u) = H(u) + s$$

where H is the usual harmonic conjugation operator. Evidently, since in addition, $\tau(H_\tau(u)) = 0$ we see that

(1.2.19) $$sL(1) = -\tau(H(u)).$$

Thus, for H_τ to be well defined it is necessary and sufficient for $L(1) \neq 0$ or equivalently, $\tau(1) \neq \tau(0)$. If $\tau(1) \neq \tau(0)$, then we see from (1.2.18) and (1.2.19) that

$$H_\tau(u) = H(u) - \frac{\tau(H(u))}{L(1)}.$$

Hence, since H is linear, H_τ is affine and since H is continuous, H_τ will be continuous if τ is continuous.

We summarize the observations from the previous paragraph in the following lemma.

LEMMA 1.2.20. *If τ is a continuous affine functional on $Harm(\Omega)$ with $\tau(1) \neq \tau(0)$, then H_τ defined by (1.2.17) is a continuous affine mapping.*

1.3. The First Herglotz Representation

In this section we shall introduce a kernel function $\Gamma_\alpha(z)$ on the domain Ω, which we call the Herglotz kernel of the first kind. Though we introduce the kernel from the point of view of functional analysis, we also shall discuss the kernel from a number of purely function-theoretic points of view at the end of this section. The existence of the $\Gamma_\alpha(z)$ was established by Helmut Grunsky in [**Gru78**]; in [**Hei85**] Maurice Heins showed that these functions are, in fact, the extreme points of a set of normalized holomorphic functions from Ω into the open right half plane. In keeping with the semi-expository style of this memoir, we shall give a self-contained development of the Herglotz kernel of the first kind. As an added benefit, we shall derive an explicit formula for the real part of this kernel in terms of the Poisson kernel and the orthonormal basis $\{\phi_1, \ldots, \phi_m\}$ of $L^2_\mathbb{R}(\partial\Omega) \ominus L^{2,h}_\mathbb{R}(\partial\Omega)$, which will be employed later in this memoir. A Herglotz representation is given in terms of the Herglotz kernel of the first type. In Section 1.6 we shall see how this Herglotz representation gives a concrete condition for Ω^- to be a spectral set for an operator.

Recall from Section 1.2 that Ω is assumed to be a bounded connected open set in $\mathbb{C}$ whose boundary consists of $m+1$ simple analytic closed curves $\partial_0, \partial_1, \ldots, \partial_m$. Let $T_\Omega = \partial_0 \times \partial_1 \times \cdots \times \partial_m$. Thus, elements $\alpha = (\alpha_0, \alpha_1, \ldots, \alpha_m) \in T_\Omega$ correspond to choosing a point α_r from each boundary component ∂_r of Ω. A very basic fact in the function theory of multiply connected domains from which we here will get a great deal of mileage is the following lemma.

LEMMA 1.3.1. *If $\mu \in M^h_\mathbb{R}(\partial\Omega)$, $\mu \geq 0$, and $\mu \neq 0$, then $\mu(\partial_r) > 0$ for each r.*

PROOF. We argue by contradiction. Accordingly, assume that $\mu \in M^h_\mathbb{R}(\partial\Omega)$, that $\mu \geq 0$, that $\mu \neq 0$, and that $\mu(\partial_r) = 0$ for some r. Since $\partial\Omega$ is assumed to consist of simple closed analytic curves, $\mu^\wedge$ reflects across ∂_r. Specifically, there exists an open set $G \subseteq \mathbb{C}$ and a $u \in Harm_\mathbb{R}(G)$ such that $\partial_r \subseteq G$ and $u \mid G \cap \Omega = \mu^\wedge \mid G \cap \Omega$. Furthermore, since $\mu^\wedge$ has a harmonic conjugate on Ω we may assume by replacing G with a smaller singly connected neighborhood if necessary, that u has a harmonic conjugate on G. In particular, the Cauchy–Riemann equations imply that

(1.3.2) $$\int_{\partial_r} \frac{du}{dn} ds = 0.$$

On the other hand, since $\mu^\wedge > 0$ on Ω and $\mu(\partial_r) = 0$, $u > 0$ on $G \cap \Omega$ and $u \equiv 0$ on ∂_r so that

(1.3.3) $$\frac{du}{dn} \leq 0 \quad \text{on } \partial_r.$$

But (1.3.2) and (1.3.3) can both hold only if $\frac{du}{dn} \equiv 0$ on ∂_r. Summarizing, u is harmonic on G, $u \equiv 0$ on ∂_r, and $\frac{du}{dn} \equiv 0$ on ∂_r. Hence $u \equiv 0$ on G, contradicting the assumption that $\mu \neq 0$. This establishes the Lemma. □

As a corollary to Lemma 1.3.1 one is able to immediately see that there can be very few extreme directions in the wedge of positive harmonic functions that have a harmonic conjugate on Ω. If X is a real vector space and $E \subseteq X$ is a wedge we say $x \in E$ is an *extreme direction in E* if $x \neq 0$ and a decomposition $x = x_1 + x_2$ with $x_1, x_2 \in E$ can only occur when $x_1 = t_1 x$ and $x_2 = t_2 x$ for two scalars t_1 and t_2 with $t_1, t_2 \geq 0$. There is a simple and geometrically obvious relationship between this notion of extreme directions of a wedge and the more familiar notion of extreme points. If E is a wedge, let us agree to say that a linear functional ρ on X *strictly slices E* if $\rho(x) > 0$ whenever $x \in E$ and $x \neq 0$. If a linear functional ρ strictly slices a wedge E, then we can introduce the convex set $E_\rho = \{x \in E : \rho(x) = 1\}$ and it is then the case that every nonzero element x of E can be represented uniquely in the form $x = t x_1$ with $t > 0$ and $x_1 \in E_\rho$, a fact which leads immediately to the following result.

LEMMA 1.3.4. *Let X be a real vector space, let $E \subseteq X$ be a wedge and let ρ be a linear functional on X that strictly slices E. If x is an extreme direction of E, then $\rho(x)^{-1} x$ is an extreme point of E_ρ. Conversely, if x is an extreme point of E_ρ and $t > 0$, then tx is an extreme direction of E.*

It is well known that the extreme directions in the wedge of finite positive Borel measures on a compact Hausdorff space are the point masses. Lemma 1.3.1 implies in a simple way that the extreme directions in the wedge of positive measures in $M_{\mathbb{R}}^h(\partial \Omega)$ are scarcely more complex.

LEMMA 1.3.5. *Let $E = \{\mu \in M_{\mathbb{R}}^h(\partial \Omega) : \mu \geq 0\}$. If μ is an extreme direction in E, then there exist $\alpha \in T_\Omega$ and $w \in \mathbb{R}^{m+1}$ such that*

(1.3.6) $$\mu = \sum_{r=0}^{m} w_r \delta_{\alpha_r}.$$

Conversely, if μ is as in (1.3.6) and $\mu \in E$, then μ is an extreme direction in E.

PROOF. We argue by contradiction. Accordingly, fix μ, an extreme direction in E, and assume that there exists an r such that the support of μ meets ∂_r in more than one point. Thus, since $\mu \geq 0$, there exist disjoint Borel sets $\Delta_1, \Delta_2 \subseteq \partial_r$ such that $\mu(\Delta_1), \mu(\Delta_2) > 0$. For $\Delta \subseteq \partial \Omega$ a Borel set, let χ_Δ denote the characteristic function of Δ so that $\chi_\Delta \mu$ is the Borel measure on $\partial \Omega$ defined by $(\chi_\Delta \mu)(S) = \mu(\Delta \cap S)$.

Now, since $\mu \in E$, $\mu \geq 0$. Also, since μ is an extreme direction, $\mu \neq 0$. Hence by Lemma 1.3.1, $\chi_{\partial_s} \mu \neq 0$ for each $s \neq r$. Thus, the $m+2$ measures, $\chi_{\partial_s} \mu (s \neq r)$, $\chi_{\Delta_1} \mu$, $\chi_{\Delta_2} \mu$ are nonzero and have pairwise disjoint supports. It follows that if $\mathcal{M}$ denotes their linear span in $M_{\mathbb{R}}(\partial \Omega)$, then $\dim(\mathcal{M}) = m+2$. Accordingly, (1.2.9) implies that

$$\dim(\mathcal{M} \cap M_{\mathbb{R}}^h(\partial \Omega)) \geq 2,$$

so that there exists $\nu \in \mathcal{M} \cap M_{\mathbb{R}}^h(\partial\Omega)$ with ν and μ linearly independent.

We now show that μ is not an extreme direction in E, a contradiction that will establish the first assertion of the lemma. By the construction of the $m+2$ positive measures that span $\mathcal{M}$, there exist $\varepsilon > 0$ such that $\mu \pm \varepsilon\nu \geq 0$. Consequently, $\frac{1}{2}(\mu \pm \varepsilon\nu) \in E$. But

$$\mu = \frac{1}{2}(\mu + \varepsilon\nu) + \frac{1}{2}(\mu - \varepsilon\nu)$$

and μ and ν are linearly independent. Thus, μ is not an extreme direction in E and the first assertion of Lemma 1.3.5 is established.

Now assume that μ is as in (1.3.6) and $\mu \in E$. Again, we argue by contradiction. If μ is not an extreme direction, then there exists $\nu \in E$ such that $\nu \leq \mu$ and $\nu \neq t\mu$ for any $t \in [0,1]$. Let t_0 be the largest t such that $\mu - t_0\nu \geq 0$ so that there exists r such that $(\mu - t_0\nu)(\partial_r) = 0$. Since $\mu - t_0\nu \in E$, this contradicts Lemma 1.3.1 and the proof of Lemma 1.3.5 is complete. □

Lemma 1.3.5 asserts that each extreme direction in $\{\mu \in M_{\mathbb{R}}^h(\partial\Omega) : \mu \geq 0\}$ has the form (1.3.6) for some $\alpha \in T_\Omega$ and some $w \in \mathbb{R}^{m+1}$. Proposition 1.3.15 below asserts that in addition, to each $\alpha \in T_\Omega$ there exists a $w \in \mathbb{R}^{m+1}$ such that μ defined by Lemma 1.3.1 is an extreme direction. Furthermore, w is unique up to scalar multiple so that in fact there is a correspondence between extreme directions of $\{\mu \in M_{\mathbb{R}}^h(\partial\Omega) : \mu \geq 0\}$ and T_Ω. The key to the proof of Proposition 1.3.15 rests in the consideration of a certain set $U \subseteq T_\Omega$. For $\alpha \in T_\Omega$ and $w \in \mathbb{R}^{m+1}$ let $\mu_{\alpha,w} \in M_{\mathbb{R}}(\partial\Omega)$ be defined by

$$(1.3.7) \qquad \mu_{\alpha,w} = \sum_{r=0}^{m} w_r \delta_{\alpha_r}.$$

Let $U \subseteq T_\Omega$ be defined by

$$(1.3.8) \qquad U = \{\alpha \in T_\Omega : \exists_{w \in \mathbb{R}^{m+1}} w \neq 0, 0 \leq \mu_{\alpha,w} \text{ and } \mu_{\alpha,w} \in M_{\mathbb{R}}^h(\partial\Omega)\}.$$

To analyze whether a given $\mu_{\alpha,w}$ is in $M_{\mathbb{R}}^h(\partial\Omega)$ we employ (1.2.9). Thus, for $\alpha \in T_\Omega$ define an $m \times (m+1)$ matrix $A(\alpha)$ by setting

$$(1.3.9) \qquad A(\alpha) = [\phi_r(\alpha_s)]_{\substack{0 \leq s \leq m \\ 1 \leq r \leq m}}.$$

Evidently, if $w \in \mathbb{R}^{m+1}$, then

$$\int \phi_r \, d\mu_{\alpha,w} = \sum_{s=0}^{m} \phi_r(\alpha_s) w_s,$$

so that by (1.2.9),

$$(1.3.10) \qquad \mu_{\alpha,w} \in M_{\mathbb{R}}^h(\partial\Omega) \quad \text{if and only if } A(\alpha)w = 0.$$

An ingredient of the proof of Proposition 1.3.15 will be the following lemma.

LEMMA 1.3.11. *If $\alpha \in U$, then $\operatorname{rank} A(\alpha) = m$.*

PROOF. We argue by contradiction. Thus, assume that $\alpha \in U$ but $\operatorname{rank} A(\alpha) < m$. By the Rank–Nullity Theorem it follows that

$$(1.3.12) \qquad \dim \ker A(\alpha) \geq 2.$$

Since $\alpha \in U$, there exists $w \in \mathbb{R}^{m+1}$ with the properties that $w \neq 0$, $\mu_{\alpha,w} \geq 0$, and $\mu_{\alpha,w} \in M_{\mathbb{R}}^h(\partial\Omega)$. In particular, by (1.3.10) $w \in \ker A(\alpha)$. Using (1.3.12) choose $v \in \ker A(\alpha)$ so that v and w are linearly independent.

Now, by Lemma 1.3.1, $w_r > 0$ for each r. Hence, there exists $\lambda \in \mathbb{R}$ such that if $u = w + \lambda v$, then u has the following two properties.

(1.3.13) $$u_r \geq 0 \quad \text{for each } r.$$
(1.3.14) $$u_r = 0 \quad \text{for some } r.$$

Now, since $w, v \in \ker A(\alpha)$ so also is $u \in \ker A(\alpha)$. Hence by (1.3.10), $\mu_{\alpha,u} \in M_{\mathbb{R}}^h(\partial\Omega)$. Also, since w and v are linearly independent, $\mu_{\alpha,u} \neq 0$. Finally, the linear independence of α and v, and (1.3.13) imply that $\mu_{\alpha,u} \neq 0$. Hence by Lemma 1.3.1,
$$u_r = \mu_{\alpha,u}(\partial_r) > 0$$
for each r. This contradiction to (1.3.14) establishes Lemma 1.3.11 □

PROPOSITION 1.3.15. *Let $E = \{\mu \in M_{\mathbb{R}}^h(\partial\Omega) : \mu \geq 0\}$. If $\alpha \in T_\Omega$, then there exists $w \in \mathbb{R}^{m+1}$ such that $\mu_{\alpha,w} \in E$. Furthermore, if w_1 and w_2 are nonzero elements of $\mathbb{R}^{m+1}$ and both μ_{α,w_1} and μ_{α,w_2} are in E, then there exists $t > 0$ such that $w_2 = tw_1$.*

PROOF. Recalling the set U defined in (1.3.8), note that the first assertion of the proposition will follow if we show that $U = T_\Omega$. Noting that T_Ω is connected, that $U = T_\Omega$ will follow if we can establish that U is both open and closed and that U is nonempty.

We first show that U is closed. Accordingly, let $\alpha_i \in U$ with $\alpha_i \to \alpha$. Pick $w \in \mathbb{R}^{m+1}$ with $\mu_{\alpha_i, w_i} \in M_{\mathbb{R}}^h(\partial\Omega)$, $u_{\alpha_i, w_i} \geq 0$, and $\|w_i\| = 1$. By compactness of the unit sphere in $\mathbb{R}^{n+1}$, there exists $w \in \mathbb{R}^{n+1}$ and a subsequence $\{w_{i_j}\}$ such that $w_{i_j} \to w$ as $j \to \infty$. Since $\mu_{\alpha_{i_j}, w_{i_j}} \to \mu_{\alpha, w} wk*$, it follows that $\mu_{\alpha,w} \geq 0$. Also, from (1.2.9) we see that $\mu_{\alpha,w} \in M_{\mathbb{R}}^h(\partial\Omega)$. Finally, we note that since $\|w_i\| = 1$, $\|w\| = 1$ so that in particular, $w \neq 0$. These facts imply that $\alpha \in U$ and we have shown that U is closed.

To see that U is open, fix $\alpha \in U$ and pick $w \neq 0$ so that $\mu_{\alpha,w} \geq 0$ and $\mu_{\alpha,w} \in M_{\mathbb{R}}^h(\partial\Omega)$. Now by (1.3.10), $w \in \ker A(\alpha)$. Since by Lemma 1.3.11, $\text{rank } A(\alpha) = m$ and since the entries of $A(\alpha)$ are continuous, there exists a neighborhood V of α and a function $f : V \to \mathbb{R}^{m+1}$ with the properties that f is continuous, $f(\alpha) = w$, and $f(\beta) \in \ker A(\beta)$ for all $\beta \in V$. Now, since $\alpha \in U$, Lemma 1.3.1 implies that $w_r = \mu_{\alpha,w}(\partial_r) > 0$ for each r. Hence, since f is continuous and $f(\alpha) = w$, there exists a neighborhood V_0 of α with $V_0 \subseteq V$ and such that $f(\beta)_r > 0$ for each r and for each $\beta \in V_0$. Consequently, $\mu_{\beta,f(\beta)} \geq 0$ and $\mu_{\beta,f(\beta)} \neq 0$ whenever $\beta \in V_0$. Since, in addition, (1.3.10) and the fact that $f(\beta) \in \ker A(\beta)$ for all $\beta \in V_0$ imply that $\mu_{\beta,f(\beta)} \in M_{\mathbb{R}}^h(\partial\Omega)$ for all $\beta \in V_0$, we see that $\beta \in U$ for all $\beta \in V_0$. Thus, U is open.

Finally, we claim that $U \neq \emptyset$. To see this note that $E \cap Prob(\partial\Omega)$ is a $wk*$ closed compact subset of $M_{\mathbb{R}}(\partial\Omega)$. Furthermore, by assumption, $\sigma \in Prob(\partial\Omega)$ and since $1 \in Hol(\Omega)$, $\sigma \in E$. Hence $\sigma \in E \cap Prob(\partial\Omega)$ and $E \cap Prob(\partial\Omega)$ is nonempty. By the Krein–Milman Theorem, $E \cap Prob(\partial\Omega)$ has an extreme point μ. Now, if we define ρ by $\rho(\nu) = \int d\nu$ then $E_\rho = E \cap Prob(\partial\Omega)$ and it follows from Lemma 1.3.4 that μ is an extreme direction of E. If α and w are as in Lemma 1.3.5 we see that $\alpha \in U$. Thus, $U \neq \emptyset$.

Summarizing, we have shown that U is a nonempty subset of T_Ω that is both open and closed. This proves the first assertion of the proposition.

To prove the second assertion of the proposition, suppose w_1 and w_2 are nonzero elements of $\mathbb{R}^{m+1}$ and both μ_{α,w_1} and μ_{α,w_2} are in E. By Lemma 1.3.1, $w_{1r} = \mu_{\alpha,w_1}(\partial_r) > 0$ for each r. Let

$$t = \min_r \frac{w_{2r}}{w_{1r}}$$

so that $(w_2 - tw_1)_r \geq 0$ for each r but $(w_2 - tw_1)_r = 0$ for some r. By Lemma 1.3.1, $\mu_{\alpha, w_2 - tw_1} = 0$. Hence $w_2 = tw_1$. $t > 0$ since $w_2 \neq 0$. This concludes the proof of Proposition 1.3.15. □

Before continuing we remark that Lemma 1.3.11 is subsumed into the following corollary of Proposition 1.3.15.

COROLLARY 1.3.16. $rank\, A(\alpha) = m$ for all $\alpha \in T_\Omega$.

PROOF. If $\dim \ker A(\alpha) \geq 2$, then the second assertion of Proposition 1.3.15 would not hold. □

We now wish to convert Lemma 1.3.5 and Proposition 1.3.15, which are results about extreme *directions*, into results about extreme *points*. Recall that $Prob(\partial \Omega)$ denotes the probability measures on $\partial \Omega$ and $Prob^h(\partial \Omega)$ denotes the probability measures that have a harmonic conjugate. Note that if ρ is defined on $M_\mathbb{R}(\partial \Omega)$ by $\rho(\mu) = \mu(\partial \Omega)$ and E is as in Lemma 1.3.5, then $E_\rho = Prob^h(\partial \Omega)$. Consequently, by Lemma 1.3.4, Lemma 1.3.5 implies that if μ is an extreme point of $Prob^h(\partial \Omega)$, then there exist $\alpha \in T_\Omega$ and $w \in \mathbb{R}^{m+1}$ such that $\mu = \mu_{\alpha,w}$. Also, if $\mu_{\alpha,w} \in Prob^h(\partial \Omega)$, then $\mu_{\alpha,w}$ is an extreme point of $Prob^h(\partial \Omega)$. Likewise, by Lemma 1.3.4, Proposition 1.3.15 implies that if $\alpha \in T_\Omega$, then there exists a unique $w \in \mathbb{R}^{m+1}$ such that $\mu_{\alpha,w} \in Prob^h(\Omega)$. Thus, the following theorem obtains.

THEOREM 1.3.17. *If $\alpha \in T_\Omega$, there exist unique positive weights $w_0, \ldots, w_m$ such that*

(1.3.18) $$w_0 \delta_{\alpha_0} + \cdots + w_m \delta_{\alpha_m} \in Prob^h(\partial \Omega).$$

Furthermore, the extreme points of $Prob^h(\partial \Omega)$ consist precisely of the measures of the form given in (1.3.18).

We now are able to formulate and prove our first Herglotz representation.

DEFINITION 1.3.19. For each r define $w_r : T_\Omega \to \mathbb{R}^+$ by requiring that $\sum_r w_r(\alpha) \delta_{\alpha_r} \in Prob^h(\partial \Omega)$. Let $w : T_\Omega \to \mathbb{R}^{m+1}$ be defined by $w(\alpha) = (w_r(\alpha))$.

Evidently, Theorem 1.3.17 guarantees that the w_r are well defined. In addition, the w_r have whatever regularity the ϕ_r have. Specifically, the following lemma obtains.

LEMMA 1.3.20. *For each $\alpha \in T_\Omega$ let the $(m+1) \times (m+1)$ matrix $B(\alpha)$ be defined by $B(\alpha)_{rs} = \phi_r(\alpha_s)$ if $1 \leq r \leq m$ and $0 \leq s \leq m$ and by $B(\alpha)_{rs} = 1$ if $r = 0$ and $0 \leq s \leq m$. Let $e_0 \in \mathbb{R}^{m+1}$ be defined by $e_{0r} = 1$ if $r = 0$ and $e_{0r} = 0$ if $1 \leq r \leq m$. For each $\alpha \in T_\Omega$, $B(\alpha)$ is invertible and*

(1.3.21) $$w(\alpha) = B(\alpha)^{-1} e_0.$$

In particular, if ϕ_r is continuous (resp. k times differentiable, infinitely differentiable, real analytic) for each r, then w is continuous (resp. k times differentiable, infinitely differentiable, real analytic).

PROOF. Observe that if the first row of $B(\alpha)$ is deleted then one obtains the matrix $A(\alpha)$ of (1.3.9). Thus, since $\mu_{\alpha,w(\alpha)} \in Prob^h(\partial\Omega) \subseteq M_\mathbb{R}^h(\partial\Omega)$ we see via (1.3.10) that $B(\alpha)w(\alpha) = \lambda e_0$ for some $\lambda \in \mathbb{R}$. On the other hand, since $\mu_{\alpha,w(\alpha)} \in Prob^h(\partial\Omega) \subseteq Prob(\partial\Omega)$, the fact that the first row of $B(\alpha)$ consists entirely of 1's implies that

$$\begin{aligned} \lambda &= \langle \lambda e_0, e_0 \rangle \\ &= \langle B(\alpha)w(\alpha), e_0 \rangle \\ &= \Sigma w_r(\alpha) \\ &= \mu_{\alpha,w(\alpha)}(\partial\Omega) \\ &= 1. \end{aligned}$$

Thus, (1.3.21) will follow if we can show that $B(\alpha)$ is invertible. Also, from the determinant formula for the inverse of a matrix it is clear that if $B(\alpha)$ is invertible, then w inherits whatever regularity the ϕ_r's have.

To see that $B(\alpha)$ is invertible let $v \in \mathbb{R}^{m+1}$ and assume that $B(\alpha)v = 0$. Thus $\Sigma v_r = 0$ and $A(\alpha)v = 0$. By Lemma 1.3.11, there exists $\lambda \in \mathbb{R}$ such that $v = \lambda w$. Hence since $\Sigma w_r = 1$, $0 = \Sigma v_r = \lambda \Sigma w_r = \lambda$ and we see that in fact, $v = 0$. This shows that $B(\alpha)$ is invertible and concludes the proof of Lemma 1.3.20. □

Our next definition introduces the class of functions on Ω to which our Herglotz representations of the first and second kind will apply.

DEFINITION 1.3.22. Define $C \subseteq Hol(\Omega)$ to consist of the set of all functions $h \in Hol(\Omega)$ such that

(1.3.23)
$$\operatorname{Re} h \geq 0,$$
$$\sigma(\operatorname{Re} h) = 1,$$

and

$$\tau(\operatorname{Im} h) = 0.$$

Before continuing, we remark that we have in our notation suppressed the dependence of C on σ and τ. We shall have quite a bit to say about C. The following proposition summarizes the facts immediately available to us.

PROPOSITION 1.3.24. C is a compact convex subset of $H(\Omega)$. $\mu \mapsto \mu^\wedge + iH_\tau(\mu^\wedge)$ is an affine homeomorphism of $(Prob^h(\partial\Omega), wk*)$ onto C.

PROOF. Define $L : Prob^h(\partial\Omega) \to C$ by the formula,

$$L(\mu) = \mu^\wedge + iH_\tau(\mu^\wedge).$$

By Lemmas 1.2.12 and 1.2.20, L is 1-1, affine, and continuous. Since L is affine and $Prob^h(\partial\Omega)$ is convex, $C = L(Prob^h(\partial\Omega))$ is convex. Since L is continuous and $(Prob^h(\partial\Omega), wk*)$ is compact, $C = L(Prob^h(\partial\Omega))$ is compact and L is a homeomorphism.

Since $C \subseteq Hol(\Omega)$ is convex and compact and since $Hol(\Omega)$ is a locally convex space, *a priori*, C is equipped with a lot of extreme points. Furthermore, if L is the affine homeomorphism of Proposition 1.3.24, h is an extreme point of C if and only if $h = L(\mu)$ for some extreme point μ of $Prob^h(\Omega)$. □

DEFINITION 1.3.25. For $\alpha \in T_\Omega$ define $\Gamma_\alpha \in Hol(\Omega)$ by requiring
$$Re\, \Gamma_\alpha = \Sigma w_r(\alpha) \delta_{\alpha_r}^\wedge,$$
and
$$\tau(\operatorname{Im} \Gamma_\alpha) = 0.$$
We refer to $\Gamma_\alpha(z)$ as the *Herglotz kernel of the first kind*.

We remark that just as with C, we have in our notation suppressed the dependence of Γ_α on σ and τ. Also, notice that as an immediate consequence of Definition 1.3.19, Theorem 1.3.17, Definition 1.3.22, and Proposition 1.3.24 we have the following theorem.

THEOREM 1.3.26. *h is an extreme point of C if and only if there exists $\alpha \in T_\Omega$ such that $h = \Gamma_\alpha$.*

Finally, we are poised to state and quickly prove our first Herglotz representation theorem for functions of positive real part on multiply connected domains.

THEOREM 1.3.27. *$h \in C$ if and only if there exists $\mu \in Prob(T_\Omega)$ such that*
$$h = \int \Gamma_\alpha \, d\mu(\alpha).$$

PROOF. We first verify that $\int \Gamma_\alpha \, d\mu(\alpha)$ makes sense. By (1.2.4),
$$\partial\Omega \ni \lambda \mapsto \delta_\lambda^\wedge \in Harm(\Omega)$$
is continuous. Hence by Lemma 1.3.20,
$$T_\Omega \ni \alpha \mapsto \mu_{\alpha,w(\alpha)}^\wedge \in Harm^h(\Omega)$$
is continuous. Hence by Lemma 1.2.20,
$$T_\Omega \ni \alpha \mapsto \Gamma_\alpha \in Hol(\Omega)$$
is continuous. Consequently, $\int \Gamma_\alpha \, d\mu(\alpha)$ is well defined either as a scalar integral, $\int \Gamma_\alpha(z) \, d\mu(\alpha)$, for each fixed $z \in \Omega$ or as the $Hol(\Omega)$-valued integral of the $Hol(\Omega)$-valued continuous function Γ_α. In addition, if we define $L : Prob(T_\Omega) \to Hol(\Omega)$ by
$$L(\mu) = \int \Gamma_\alpha \, d\mu$$
then L is continuous if $Prob(T_\Omega)$ is equipped with the $wk*$ topology.

Now let $E = \{\int \Gamma_\alpha \, d\mu : \mu \in Prob(T_\Omega)\}$. Evidently, we need to show that $E = C$. Since $L(\delta_\alpha) = \Gamma_\alpha \in C$ and since the convex hull of $\{\delta_\alpha : \alpha \in T_\Omega\}$ is $wk*$ dense in $Prob(T_\Omega)$ we see immediately from the facts that L is affine and continuous that $E = L(Prob(T_\Omega)) \subseteq C$. Conversely, note that $\Gamma_\alpha = L(\delta_\alpha) \in E$ so that by Theorem 1.3.26, the extreme points of C lie in E. But Proposition 1.3.24 implies that C is convex and compact and the linearity and continuity of L imply that E is convex and compact. Hence the Krein–Milman Theorem implies that $C \subseteq E$. This concludes the proof of Theorem 1.3.27. □

We conclude this section with the function-theoretic properties of the Herglotz kernel of the first type, viewed as a mapping $z \mapsto \Gamma_\alpha(z)$ from Ω into the right open half plane, $\mathbb{H}$, for each $\alpha \in T_\Omega$. Our analysis will be based on the following

1.3. THE FIRST HERGLOTZ REPRESENTATION

well-known properties of the Poisson kernel (recall our standing assumption that $\partial\Omega$ is analytic).

(1.3.28)   For every $\lambda \in \partial\Omega$, there exists a neighborhood U of Ω^- such that the map $z \mapsto P_z^\sigma(\lambda)$ extends harmonically, by Schwartz reflection, to $U\setminus\{\lambda\}$.

(1.3.29)   For every $\lambda \in \partial\Omega, z \mapsto P_z^\sigma(\lambda)$, is strictly positive on Ω and vanishes identically on $\partial\Omega\setminus\{\lambda\}$.

(1.3.30)   For every $\lambda \in \partial\Omega, z \mapsto P_z^\sigma(\lambda) - \operatorname{Re}\left(\dfrac{e^{i\theta}}{\pi \frac{d\sigma}{ds}(\lambda)} \dfrac{1}{z-\lambda}\right)$ extends harmonically to a neighborhood of λ, where θ is the direction of the inward normal to $\partial\Omega$ at λ, and s is arc length measure on $\partial\Omega$.

A thorough development of these properties can be found in Grunsky's book [**Gru78**], although $P_z^\sigma(\lambda)$ is scaled differently than Grunsky's kernel $p(z, \lambda)$ so as to have the reproducing property (1.2.2). Properties (1.3.28)–(1.3.30) translate into the following function-theoretic characterization of the Herglotz kernel of the first type.

PROPOSITION 1.3.31. *Let $f \in Hol(\Omega)$. The following are equivalent.*
 (1) *There exist $\alpha \in T_\Omega$, $a > 0$, and $b \in \mathbb{R}$ such that $f = a\Gamma_\alpha + ib$.*
 (2) *f extends meromorphically to a neighborhood $\mathcal{N}$ of Ω^- and, so extended, possesses the following properties.*

(1.3.32) f *maps Ω onto $\mathbb{H}$.*

(1.3.33)   *For all $r = 0, \ldots, m$, f has exactly one pole on ∂_r and maps ∂_r 1-1 and onto the extended imaginary axis.*

In addition, if f satisfies the above equivalent criteria then, when extended to $\mathcal{N}$, its poles on $\partial\Omega$ are the points $z = \alpha_0, \ldots, \alpha_m$, and these poles determine f uniquely up to a positive scaling and an imaginary translations.

PROOF. (1) $\Rightarrow$ (2) and the assertion that the poles of Γ_α are $\partial\Omega$ are the points $\alpha_0, \ldots, \alpha_m$ are straightforward consequences of the definition of Γ_α, formula (1.2.16) and properties (1.3.28)–(1.3.30) of the Poisson kernel.

To prove that (2) $\Rightarrow$ (1), assume (2) and view f as having been extended meromorphically to $\mathcal{N}$. Let $\alpha_0, \ldots, \alpha_m$ denote the poles of f on $\partial_0, \ldots, \partial_m$ respectively. By (1.3.32) and (1.3.33), for each $r = 0, \ldots, m$ α_r is a simple pole; moreover there exists $t_r > 0$ such that
$$f(z) - \frac{t_r e^{i\theta_r}}{z - \alpha_r}$$
has a removable singularity at α_r, where θ_r is the direction of the inward normal to $\partial\Omega$ at α_r. Thus,

(1.3.34) $$f(z) - \sum_{r=0}^{m} \frac{t_r e^{i\theta_r}}{z - \alpha_r} \text{ extends holomorphically past } \partial\Omega.$$

Define
$$\omega_r = t_r \pi \frac{d\sigma}{ds}(\alpha_r) \qquad r = 0, \ldots, m$$
and
$$u(z) = \operatorname{Re} f(z) - \sum_{r=0}^{m} \omega_r P_z^{\sigma}(\alpha_r).$$

Comparing (1.3.34) and (1.3.30), we conclude that u extends harmonically to a neighborhood of Ω^-. But by (1.3.29) and (1.3.33), u vanishes identically on $\partial\Omega \setminus \{\alpha_0, \ldots, \alpha_m\}$, hence, when extended, it vanishes on all of $\partial\Omega$. It follows that $u \equiv 0$, thus
$$\operatorname{Re} f(z) = \sum_{r=0}^{m} \omega_r P_z(\alpha_r) = \sum_{r=0}^{m} \omega_r \delta_{\alpha_r}^{\wedge}(z).$$

Recalling the definition of Γ_α (1.3.21) and invoking Proposition 1.3.15, we conclude that there exists $a \geq 0$ such that
$$\operatorname{Re} f = a \operatorname{Re} \Gamma_\alpha.$$

The harmonic conjugates of $\operatorname{Re} f$ and $a \operatorname{Re} \Gamma_\alpha$ can hence differ by at most an imaginary constant, establishing (1).

In summary, we have proven the equivalence of (1) and (2), and that the poles of Γ_α on $\partial\Omega$ are precisely the points $\alpha_0, \ldots, \alpha_m$. The assertion that any $f \in Hol(\Omega)$ satisfying (1) and (2) is determined by its poles on $\partial\Omega$ uniquely up to positive scaling and an imaginary translation is evident from the fact that Γ_α is uniquely determined by α. This completes the proof of Proposition 1.3.31. $\square$

Let $\mathcal{M}$ denote the set of $m+1$-valent holomorphic functions from Ω onto $\mathbb{H}$. Using Proposition 1.3.31, it is straightforward to show that

(1.3.35) $$\mathcal{M} = \{a\Gamma_\alpha + ib : \alpha \in T_\Omega, a > 0, b \in \mathbb{R}\}.$$

From this parameterization of $\mathcal{M}$, together with (1.3.23) and Theorem 1.3.26, we recover Forelli's result that the normalized $m+1$-valent holomorphic functions from Ω into $\mathbb{H}$ are precisely the extreme points of C [**For79**]. We remark that by applying the argument principle one can show that $\mathcal{M}$ is, in fact, the set of single-valued holomorphic functions from Ω onto $\mathbb{H}$ of minimal constant valence. Readers acquainted with the theory of Blashke products on multiply connected domains will recognize that the set
$$\mathcal{M}' = \left\{ \frac{f-1}{f+1} : f \in \mathcal{M} \right\}$$
is comprised of all single-valued Blashke products on Ω of order $m+1$ (or, equivalently, of minimal order). From (1.3.35) we immediately see that such Blashke products are parameterized continuously and uniquely by $m+3$ real parameters; moreover, when viewed as a subset of the algebra of homomorphic functions on Ω which are continuous up to $\partial\Omega$, equipped with the supremum norm, $\mathcal{M}'$ is homeomorphic to $T_\Omega \times \mathbb{R}^+ \times \mathbb{R}$.

Perhaps the most well known context in which functions in $\mathcal{M}'$ arise is given by the extremal problem

(1.3.36) $$\sup\{f'(z_0) : f : \Omega \to \mathbb{D} \text{ is holomorphic, } f'(z_0) > 0\}$$

where $z_0 \in \Omega$. For each $z_0 \in \Omega$, a unique solution F_{z_0}, known as the Ahlfors function at z_0, exists and can be shown to be a member of $\mathcal{M}'$ satisfying $F_{z_0}(z_0) = 0$ (see, for example [**Fis83**, Chapter 5]).

In [**FK99**] it was shown that the converse fails to hold—that is, not every member of $\mathcal{M}'$ arises from (1.3.36)—a fact that we can now see from a simple parameter count. The set of all Ahlfors functions on Ω is parameterized continuously by Ω via $z_0 \mapsto F_{z_0}$, although this parameterization need not be 1-1 (since the same Ahlfors function may solve (1.3.36) for two distinct choices of z_0). In any case the set of all Ahlfors functions is parameterized continuously by two real numbers. Taking into account the fact that $Aug(\mathbb{D})$ is parameterized continuously and uniquely by three real numbers and $\mathcal{M}'$ by $m+3$ real numbers, we conclude that for $m \geq 3$, there exist functions $f \in \mathcal{M}'$ which cannot be expressed in the form $f = \tau \circ A$ where $\tau \in Aut(\mathbb{D})$ and A is an Ahlfors function on Ω.

1.4. The Second Herglotz Representation

In this section we shall introduce a kernel function $H_\lambda(z)$ on the domain Ω, which we shall call the Herglotz kernel of the second kind. As in the previous section a Herglotz representation will be given in terms of the kernel. Our discussion will be much more straightforward than in the previous section as the nature of the representation is much more transparent. In addition, we describe here the kernel purely from a function-theoretic point of view reserving to the next section a functional-analytic description.

Fix $\lambda \in \partial\Omega$. If $m > 0$, then by Lemma 1.3.1, $\delta_\lambda \notin M_{\mathbb{R}}^h(\partial\Omega)$. On the other hand, if we consider the measure ν_λ defined by

$$(1.4.1) \qquad \nu_\lambda = \delta_\lambda - \sum_{r=1}^{m} \phi_r(\lambda)\phi_r \sigma,$$

then because $\{\phi_r : 1 \leq r \leq m\}$ is an orthonormal set in $L_{\mathbb{R}}^2(d\sigma)$,

$$\int \phi_s \, d\nu_\lambda = \phi_s(\lambda) - \sum_{r=1}^{m} \phi_r(\lambda) \int \phi_s \phi_r \, d\sigma$$
$$= \phi_s(\lambda) - \phi_s(\lambda)$$
$$= 0$$

for each s. Hence (1.2.9) implies that $\nu_\lambda \in M_{\mathbb{R}}^h(\partial\Omega)$.

DEFINITION 1.4.2. For $\lambda \in \partial\Omega$ define $H_\lambda \in Hol(\Omega)$ by requiring that

$$Re \, H_\lambda = \nu_\lambda^\wedge,$$

and

$$\tau(H_\lambda) = 0.$$

We first note that since $\{\phi_r\}$ is an orthonormal basis for $L_{\mathbb{R}}^{2,h}(\sigma)^\perp$, that in particular $\phi_r \perp 1$, i.e. $\int \phi_r \, d\sigma = 0$. Hence $\int d\nu_\lambda = 1$. On the other hand, $\nu_\lambda \notin Prob^h(\partial\Omega)$. To see this note that if $\nu_\lambda \geq 0$, then $-\sum \phi_r(\lambda)\phi_r\sigma \geq 0$. Since $\int \phi_r \, d\sigma = 0$, this implies that $\sum \phi_r(\lambda)\phi_r = 0$ in $L_{\mathbb{R}}^2(\sigma)$. Since $\{\phi_r\}$ is linearly independent, we deduce that $\phi_r(\lambda) = 0$ for each r. But then $\delta_\lambda = \nu_\lambda \in M_{\mathbb{R}}^h(\partial\Omega)$, a contradiction. Thus, though H_λ satisfies the normalizing conditions on C ($Re \, H_\lambda = \mu^\wedge$ for μ with $\int d\mu = 1$ and $\tau(Im \, H_\lambda) = 0$), nevertheless since $Re \, H_\lambda \not\geq 0$, $H_\lambda \notin C$.

Now, reasoning similar to that at the beginning of the proof of Theorem 1.3.27 shows that
$$\partial\Omega \ni \lambda \mapsto H_\lambda \in Hol(\Omega)$$
is continuous. Furthermore, if we define a map $L : M_\mathbb{R}(\partial\Omega) \to Hol(\Omega)$ by the formula
$$L(\mu) = \int H_\lambda \, d\mu(\lambda),$$
then L is an affine continuous map if $M_\mathbb{R}(\partial\Omega)$ is equipped with the $wk*$ topology.

Since $H_\lambda \notin C$, in general, $\int H_\lambda \, d\mu \notin C$. However, we claim that if $\mu \in Prob^h(\partial\Omega)$, then $\int H_\lambda \, d\mu \in C$. To see this, fix $\mu \in M_\mathbb{R}(\partial\Omega)$ and observe that if $z \in \Omega$, then
$$Re \int H_\lambda(z) \, d\mu(\lambda) = \int Re \, H_\lambda(z) \, d\mu(\lambda)$$
$$= \int \nu_\lambda^\wedge(z) \, d\mu(\lambda)$$
$$= \int \delta_\lambda^\wedge(z) \, d\mu(\lambda) - \sum_{r=1}^{m} \left(\int \phi_r \, d\mu \right) (\phi_r \sigma)^\wedge(z).$$

But (1.2.4) and (1.2.3) imply that
$$\int \delta_\lambda^\wedge(z) \, d\mu(\lambda) = \int P_z^\sigma \, d\mu$$
$$= \mu^\wedge(z).$$

Hence, if $\mu \in M_\mathbb{R}(\partial\Omega)$, then

(1.4.3) $$Re \int H_\lambda \, d\mu = \mu^\wedge - \sum_{r=1}^{m} \left(\int \phi_r \, d\mu \right) (\phi_r \sigma)^\wedge.$$

Now, if in addition, $\mu \in M_\mathbb{R}^h(\partial\Omega)$, then $\int \phi_r \, d\mu = 0$ for each r. Hence if $\mu \in M_\mathbb{R}^h(\partial\Omega)$,

(1.4.4) $$Re \int H_\lambda \, d\mu = \mu^\wedge.$$

From (1.4.4)) and the definition of C we see that if $\mu \in Prob^h(\partial\Omega)$, then $\int H_\lambda \, d\mu$ will be in C provided $\tau(Im \int H_\lambda \, d\mu) = 0$. But since by definition, $\tau(Im \, H_\lambda) = 0$ for each $\lambda \in \partial\Omega$, $\tau(Im \, (\int H_\lambda \, d\mu)) = \int \tau(Im \, H_\lambda) \, d\mu = 0$. Hence $\int H_\lambda \, d\mu \in C$ whenever $\mu \in Prob^h(\partial\Omega)$.

THEOREM 1.4.5. $h \in C$ if and only if there exists $\mu \in Prob^h(\partial\Omega)$ such that

(1.4.6) $$h = \int H_\lambda \, d\mu(\lambda).$$

Furthermore, if $h \in C$ and (1.4.6) holds, then

(1.4.7) $\qquad Re \, h \circ \tau_\ell \sigma \to \mu \qquad$ in the $wk *$ topology.

In particular, if $h \in C$ then (1.4.6) holds for a unique $\mu \in Prob^h(\partial\Omega)$.

PROOF. Let $C_0 = \{h \in Hol(\Omega) : (1.4.6) \text{ holds for some } \mu \in Prob^h(\partial\Omega)\}$ so that the first assertion of Theorem 1.4.5 is equivalent to the assertion that $C_0 = C$. By the remarks leading up to the statement of the theorem, $C_0 \subseteq C$. Conversely, if $h \in C$, then by (1.2.11) there exists $\mu \in Prob^h(\partial\Omega)$ such that (1.4.7) holds. It follows from (1.4.4) that (1.4.6) holds, i.e. $h \in C_0$. Thus, $C_0 = C$. The other assertions of Theorem 1.4.5 are straightforward to verify as well. □

In the literature one can find other examples of kernels $k : \Omega \times \partial\Omega \to \mathbb{C}$ with the reproducing property (1.4.6)—i.e., for every $f \in C$,

$$(1.4.8) \qquad f(z) = \int_{\partial\Omega} k(z,\lambda)\, d\mu(\lambda),$$

where μ is the boundary value of $Re\, f$ [**CW67, FK99, Kha84, Zmo58**]. These are generally referred to as Schwartz kernels. For example, in [**CW67**], Coifman and Weiss construct a Schwartz kernel whose real part is given by

$$P_z^\sigma(\lambda) - \sum_{r=1}^{m} \Lambda_r(\lambda)\omega_r(z),$$

where σ is arc length measure, $\omega_1, \ldots, \omega_m$ are harmonic measures for $\partial_1, \ldots, \partial_m$ respectively, and $\Lambda_1, \ldots, \Lambda_m$ are continuous functions on $\partial\Omega$ which can be expressed in terms of the normal derivatives of $\omega_1, \ldots, \omega_m$. In Chapter 4 we shall provide a characterization of the set of all Schwartz kernels on a multiply-connected domain, as well as a general method for constructing the real part of each such kernel from the Poisson kernel.

In summary, from the point of the reproducing property (1.4.8), there is nothing intrinsic about the choice $k(z,\lambda) = H_\lambda(z)$, the Herglotz kernel of the first type. However, in the next section we shall see that $H_\lambda(z)$ is, in fact, intrinsic from a functional analytic perspective, and that this perspective leads naturally to a third generalization of the Herglotz representation theorem.

We conclude this section by pointing out a very transparent relationship between the Herglotz kernel of the first kind and the Herglotz kernel of the second kind which follows directly from Definition 1.3.25, Definition 1.4.2 and formula (1.3.21):

$$(1.4.9) \qquad \Gamma_\alpha = \sum_{r=0}^{m} w_r(\alpha) H_{\alpha_r} \qquad \text{for all } \alpha \in T_\Omega.$$

This formula has important operator theoretic ramifications, as we shall see in Section 1.7, and in Chapter 2.

1.5. The Third Herglotz Representation

In this section we discuss yet a third generalization of the Herglotz Representation to multiply connected domains. As a by-product of our analysis we obtain a functional-analytic interpretation of the Herglotz kernel of the second kind discussed in Section 1.4.

Our first result, Proposition 1.5.2 below, will give a characterization of $^\perp(Prob^h(\partial\Omega))$. Its proof requires the following lemma.

LEMMA 1.5.1. *If $a_1, \ldots, a_m \in \mathbb{R}$ and $a_1\phi_1 + \cdots + a_m\phi_m \geq 0$, then $a_r = 0$ for each r.*

PROOF. Since $\phi_r \in L^{2,h}(\sigma)^\perp$,
$$\int \phi_r \, d\sigma = \langle 1, \phi_r \rangle$$
$$= 0.$$

Hence if $a_r \in \mathbb{R}$, then $\int \Sigma a_r \phi_r \, d\sigma = 0$. If, in addition $\Sigma a_r \phi_r \geq 0$, we deduce that $\Sigma a_r \phi_r = 0$ in $L^2(\sigma)$. But $\{\phi_r\}$ is a linearly independent set in $L^2(\sigma)$. Hence if $\Sigma a_r \phi_r \geq 0$, then $a_r = 0$ for each r and Lemma 1.5.1 is proved. □

PROPOSITION 1.5.2. Fix $v \in C_\mathbb{R}(\partial\Omega)$.
$$\int v \, d\nu \geq 0 \qquad \text{for all } \nu \in Prob^h(\partial\Omega)$$
if and only if there exist constants $a_1, \ldots, a_m \in \mathbb{R}$ such that
$$v(\lambda) + \sum_{r=1}^m a_r \phi_r(\lambda) \geq 0 \qquad \text{for all } \lambda \in \partial\Omega.$$

PROOF. Define E and $\mathcal{M}$ in $C_\mathbb{R}(\partial\Omega)$ by $E = \{v \in C_\mathbb{R}(\partial\Omega) : v \geq 0\}$ and $\mathcal{M} = \text{span}\{\phi_1, \ldots, \phi_m\}$. With this notation the proposition asserts that
$$E + \mathcal{M} = {}^\perp(Prob^h(\partial\Omega)).$$

Also, obviously,
$$(E + \mathcal{M})^\perp = \{\mu \in M_\mathbb{R}^h(\partial\Omega) : \mu \geq 0\}.$$
Since ${}^\perp\{\mu \in M_\mathbb{R}(\partial\Omega) : \mu \geq 0\} = {}^\perp(Prob^h(\partial\Omega))$, the proposition will follow from (1.2.7) if we can show that $E + \mathcal{M}$ is closed.

To see that $E + \mathcal{M}$ is closed, suppose that $v_j + \psi_j \to u$ in $C_\mathbb{R}(\partial\Omega)$, $v_j \in E$, and $\psi_j \in \mathcal{M}$. We claim that $\|\psi_j\|$ is bounded. If not, use the fact that $\mathcal{M}$ is finite dimensional to choose subsequence $\{\psi_{j_\ell}\}$ such that
$$\frac{\psi_{j_\ell}}{\|\psi_{j_\ell}\|} \to f \neq 0 \qquad \text{and } \|\psi_{j_\ell}\| \to \infty.$$
Since $v_j + \psi_j \to u$,
$$\frac{v_{j_\ell}}{\|\psi_{j_\ell}\|} \to -f.$$
Thus, $f \neq 0$, $f \in \mathcal{M}$, and $-f \in E$, a contradiction to Lemma 1.5.1. Hence $\|\psi_j\|$ is bounded.

Now since $\|\psi_j\|$ is bounded and $\psi_j \in \mathcal{M}$, a finite dimensional space, there exists a subsequence $\{\psi_{j_\ell}\}$ such that $\psi_{j_\ell} \to \psi \in \mathcal{M}$. Since $v_j + \psi_j \to u$ it follows that there exists $v \in E$ such that $v_{j_\ell} \to v$. Since $u = v + \psi$ we deduce that $u \in E + \mathcal{M}$ and we have proven that $E + \mathcal{M}$ is closed. This completes the proof of Proposition 1.5.2. □

Proposition 1.5.2 has the following elegant corollary. In addition to being an interesting piece of function theory, in Section 1.7 the result will play an important role in devising an effective numerical check for the rational dilation of 3×3 matrices to $\partial\Omega$.

PROPOSITION 1.5.3. Fix $v \in C_\mathbb{R}(\partial\Omega)$. There exist constants $a_1, \ldots, a_m \in \mathbb{R}$ such that

(1.5.4) $\qquad v(\lambda) + a_1\phi_1(\lambda) + \cdots + a_m\phi_m(\lambda) \geq 0 \qquad \text{for all } \lambda \in \partial\Omega$

if and only if

(1.5.5) $\quad w_0(\alpha)v(\alpha_0) + \cdots + w_m(\alpha)v(\alpha_m) \geq 0 \quad$ for all $\alpha \in T_\Omega$.

PROOF. By Proposition 1.5.2, (1.5.4) holds if and only if

(1.5.6) $\quad \int v\, d\mu \geq 0 \quad$ for all $\mu \in \mathit{Prob}^h(\partial\Omega)$.

But $\mathit{Prob}^h(\partial\Omega)$ is compact convex set in $(M_\mathbb{R}(\partial\Omega), wk*)$. Hence by the Krein–Milman Theorem, (1.5.6) holds if and only if

(1.5.7) $\quad \int v\, d\mu \geq 0 \quad$ whenever μ is an extreme point of $\mathit{Prob}^h(\partial\Omega)$.

But by Theorem 1.3.17 and Definition 1.3.19, μ is an extreme point of $\mathit{Prob}^h(\partial\Omega)$ if and only if $\mu = \Sigma w_r(\alpha)\delta_{\alpha_r}$ for some $\alpha \in T_\Omega$. Hence (1.5.7) is equivalent to (1.5.5) and the proof of Proposition 1.5.3 is complete. $\square$

We turn now to a dual version of Proposition 1.5.2. Noting that $\{v \in C_\mathbb{R}^h(\partial\Omega) : v \geq 0\}$ is the continuous analog of $\{\mu \in M_\mathbb{R}^h(\partial\Omega) : \mu \geq 0\}$, we can ask for a concrete characterization of $\{v \in C_\mathbb{R}^h(\partial\Omega) : v \geq 0\}^\perp$.

PROPOSITION 1.5.8. Fix $\nu \in M_\mathbb{R}(\partial\Omega)$.

$$\int v\, d\nu \geq 0 \quad \text{for all } v \in C_\mathbb{R}^h(\partial\Omega) \text{ such that } v \geq 0$$

if and only if there exist constants $a_1, \ldots, a_m \in \mathbb{R}$ such that

$$\nu + \sum_{r=1}^m a_r \phi_r \sigma \geq 0.$$

PROOF. Define E and $\mathcal{M}$ in $M_\mathbb{R}(\partial\Omega)$ by $E = \{\nu \in M_\mathbb{R}(\partial\Omega) : \nu \geq 0\}$ and $\mathcal{M} = \mathrm{span}\{\phi_1\sigma, \ldots, \phi_m\sigma\}$. Just as in the proof of Proposition 1.5.2, Proposition 1.5.8 will follow from (1.2.8) if we can show that $E + \mathcal{M}$ is $wk*$ closed.

To see that $E + \mathcal{M}$ is $wk*$ closed, for each $r > 0$ define S_r by

$$S_r = \{\mu \in E + \mathcal{M} : \|\mu\| \leq r\}.$$

By the Krein–Smulian Theorem, $E + \mathcal{M}$ will be $wk*$ closed provided S_r is $wk*$ closed for each $r > 0$. Since $S_r \subseteq r\, \mathit{ball}(M_\mathbb{R}(\partial\Omega))$ and $r\, \mathit{ball}(M_\mathbb{R}(\partial\Omega))$ with the $wk*$ topology is metrizable, S_r will be $wk*$ closed provided S_r is $wk*$ sequentially closed.

To see that S_r is $wk*$ sequentially closed fix a sequence $\nu_j + v_j\sigma \in S_r$ with $\nu_j \in E$, $v_j\sigma \in \mathcal{M}$ and $\nu_j + v_j\sigma \xrightarrow{wk*} \mu$. Since $\nu_j + v_j\sigma \xrightarrow{wk*} \mu$, in particular,

$$\int d\nu_j + \int v_j\, d\sigma \to \int d\mu.$$

On the other hand, since $v_j\sigma \in \mathcal{M}$, $\int v_j\, d\sigma = 0$. Hence,

(1.5.9) $\quad \displaystyle\int d\nu_j \to \int d\mu.$

Recalling that $\nu_j \in E = \{\nu \in M_\mathbb{R}(\partial\Omega) : \nu \geq 0\}$ we deduce from (1.5.9) that $\{\nu_j\}$ is bounded in $M_\mathbb{R}(\partial\Omega)$. Hence by Alaoglu's Theorem there exists a subsequence $\{\nu_{j_\ell}\}$ and a $\nu \in E$ such that $\nu_{j_\ell} \xrightarrow{wk*} \nu$. Since $\nu_j + v_j\sigma \xrightarrow{wk*} \mu$ we also have that there exists $v\sigma \in \mathcal{M}$ such that $v_j\sigma \xrightarrow{wk*} v\sigma$. Thus, $\mu = \nu + v\sigma \in S_r$ and we have proved that S_r is $wk*$ closed. This completes the proof of Proposition 1.5.8. □

We now are able to formulate our third generalization of the Herglotz Theorem to multiply connected domains. Recall that $h \in C$ if and only if $\operatorname{Re} h = \nu^\wedge$ for some $\nu \in M_\mathbb{R}(\partial\Omega)$ with $\nu \geq 0$ and in addition h satisfies the normalizing conditions $\int d\nu = 1$ and $\tau(\operatorname{Im} h) = 0$. Now, if $\Omega = \mathbb{D}$, $C_\mathbb{R}(\partial\Omega) = C_\mathbb{R}^h(\partial\Omega)$. Hence, the condition in the definition of C that $\nu \geq 0$ can be replaced by the equivalent condition that

$$\int v\, d\nu \geq 0 \quad \text{whenever } v \in \{u \in C_\mathbb{R}^h(\mathbb{D}) : u \geq 0\}.$$

Thus, the following definition gives a consistent way to generalize a normalized class of analytic functions with positive real part on $\mathbb{D}$ to the more general domains Ω.

DEFINITION 1.5.10. Let $P = \{v \in C_\mathbb{R}^h(\partial\Omega) : v \geq 0\}$. Let $\mathcal{P} = \{\nu \in P^\perp \cap M_\mathbb{R}^h(\partial\Omega) : \int d\nu = 1\}$. Finally, let $C_1 = \{h \in \operatorname{Hol}(\Omega) : \operatorname{Re} h = \nu^\wedge$ for some $\nu \in \mathcal{P}$ and $\tau(\operatorname{Im} h) = 0\}$.

The following proposition, which is essentially an abstract reformulation of Proposition 1.5.8, is the key to understanding the set C_1.

PROPOSITION 1.5.11. The map $\mu \mapsto \mu - \sum_r(\int \phi_r\, d\mu)\phi_r\sigma$ is an affine homeomorphism from $(\operatorname{Prob}(\partial\Omega), wk*)$ onto $(\mathcal{P}, wk*)$.

PROOF. First suppose that $\mu \in \operatorname{Prob}(\partial\Omega)$ and $\nu = \mu - \sum_r(\int \phi_r\, d\mu)\phi_r\sigma$. We claim that $\nu \in \mathcal{P}$. By Proposition 1.5.8, $\nu \in P^\perp$. Since 1 is orthogonal to ϕ_r in $L^2(\sigma)$ for each r, $\int \phi_r\, d\sigma = 0$. Hence, $\int d\nu = \int d\mu = 1$. Finally, (1.2.9) implies that $\nu \in M_\mathbb{R}^h(\partial\Omega)$. Thus $\nu \in \mathcal{P}$.

We now claim that the map is onto. To see this fix $\nu \in \mathcal{P}$. Since $\nu \in P^\perp$, Proposition 1.5.8 implies that there exist $a_1, \ldots, a_m \in \mathbb{R}$ and $\mu \in M_\mathbb{R}(\partial\Omega)$ such that $\nu = \mu + \sum a_r \phi_r \sigma$ and $\mu \geq 0$. Since $\nu \in M_\mathbb{R}^h$, (1.2.9) implies that $a_r = -\int \phi_r\, d\mu$ for each r. Finally, since $\int d\nu = 1$ and $\int \phi_r\, d\sigma = 0$ for each r, we see that $\int d\mu = 1$ i.e. $\mu \in \operatorname{Prob}(\partial\Omega)$. Thus, $\nu = \mu - \sum(\int \phi_r\, d\mu)\phi_r\sigma$ with $\mu \in \operatorname{Prob}(\partial\Omega)$ and we see that the map is onto.

To see that the map is 1-1 suppose that $\mu_1 - \sum\left(\int \phi_r\, d\mu_1\right)\phi_r\sigma = \mu_2 - \sum\left(\int \phi_r\, d\mu_2\right)\phi_r\sigma$. Since $\{\phi_1, \ldots, \phi_m\}$ is an orthonormal basis for $L_\mathbb{R}^2(\sigma) \ominus L_\mathbb{R}^{2,h}(\sigma)$ it follows that

$$\int \phi_r\, d\mu_1 = 0 = \int \phi_r\, d\mu_2 \quad \text{for all } r$$

and

$$\int \psi\, d\mu_1 = \int \psi\, d\mu_2 \quad \text{for all } \psi \in C_\mathbb{R}^h(\partial\Omega).$$

Thus, since $C_\mathbb{R}^h(\partial\Omega) + \operatorname{span}\{\phi_1, \ldots, \phi_m\} = C_\mathbb{R}(\partial\Omega)$, $\mu_1 = \mu_2$ and we have shown that the map is 1-1.

Finally, we remark that the map is obviously affine. Also, since it is clearly $wk*$ continuous and since $(\operatorname{Prob}(\partial\Omega), wk*)$ is compact, the map is a homeomorphism. This concludes the proof of Proposition 1.5.11. □

We now are able to quickly derive a number of facts about the set C_1 introduced in Definition 1.5.10.

1.5. THE THIRD HERGLOTZ REPRESENTATION

PROPOSITION 1.5.12. *If the map $L : Prob(\partial\Omega) \to C_1$ is defined by requiring that*
$$\operatorname{Re} L(\mu) = \left(\mu - \sum_r \left(\int \phi_r \, d\mu\right) \phi_r \sigma\right)^{\wedge}$$
and
$$\tau(\operatorname{Im} L(\mu)) = 0,$$
then L is an affine homeomorphism. In particular, C_1 is a convex compact subset of $Hol(\Omega)$.

PROOF. The proposition follows immediately from Proposition 1.5.11 and Lemma 1.2.20. □

Since C_1 is a convex compact subset of the locally convex topological vector space $Hol(\Omega)$, the Krein–Milman Theorem guarantees that C_1 has a plentiful supply of extreme points. The following theorem identifies these extreme points and in addition gives the functional analytic interpretation of the Herglotz kernel of the second kind promised in the introduction and in Section 1.4.

THEOREM 1.5.13. *Fix $h \in Hol(\Omega)$. h is an extreme point of C_1 if and only if $h = H_\lambda$ for some $\lambda \in \partial\Omega$.*

PROOF. By Proposition 1.5.12 h is an extreme point of C_1 if and only if $h = L(\mu)$ for some extreme point μ of $Prob(\partial\Omega)$ (here, L is as defined in Proposition 1.5.12). But μ is an extreme point of $Prob(\partial\Omega)$ if and only if $\mu = \delta_\lambda$ for some $\lambda \in \partial\Omega$. Finally, comparing (1.4.1) and Definition 1.4.2 with the definition of L given in Proposition 1.5.12 allows one to see quickly that $L(\delta_\lambda) = H_\lambda$, and we are done. □

Finally, we state and prove our their Herglotz representation.

THEOREM 1.5.14. *$h \in C_1$ if and only if there exists $\mu \in Prob(\partial\Omega)$ such that*
$$(1.5.15) \qquad h = \int H_\lambda \, d\mu.$$

Furthermore, if $h \in C_1$, then (1.5.15) holds for a unique $\mu \in Prob(\partial\Omega)$ and
$$(\operatorname{Re} h \circ \tau_\ell)\sigma \to \mu - \sum \left(\int \phi_r \, d\mu\right) \phi_r \sigma \quad \text{in the } wk* \text{ topology.}$$

PROOF. Observe that if L is as in Proposition 1.5.12, then (1.4.3) implies that
$$\operatorname{Re}\left(\int H_\lambda \, d\mu\right) = \operatorname{Re} L(\mu).$$
Also,
$$\tau\left(\operatorname{Im} \int H_\lambda \, d\mu\right) = \int \tau(\operatorname{Im} H_\lambda) \, d\mu$$
$$= \int 0 \, d\mu$$
$$= 0$$
$$= \tau(\operatorname{Im} L(\mu)).$$
Hence, if $\mu \in Prob(\partial\Omega)$, then $\int H_\lambda \, d\mu = L(\mu)$ and we see that Theorem 4.9 follows immediately from Proposition 1.5.12 and (1.2.11). □

1.6. The Herglotz Representations and Operator Theory

In this section, following the notation introduced in Section 1.1, we shall let $\mathcal{S}$ denote the collection of operators T such that Ω^- is a spectral set for T. Likewise, $\mathcal{D}$ will denote the collection of operators T that have a rational dilation $\partial\Omega$. Recall that if $T \in \mathcal{L}(\mathcal{H})$, $\sigma(T) \subset \Omega$, and $f \in Hol(\Omega)$, then $f(T) \in \mathcal{L}(\mathcal{H})$ can be defined by the formula

$$(1.6.1) \qquad f(T) = \frac{1}{2\pi i} \int_\gamma \frac{f(\lambda)}{\lambda - T} \, d\lambda.$$

In (1.6.1), γ is a system of oriented curves in Ω with the properties that $\sigma(T)$ lies inside γ and $\mathbb{C}\backslash\Omega$ lies outside γ. Furthermore, the map $f \mapsto f(T)$ from $Hol(\Omega)$ into $\mathcal{L}(\mathcal{H})$ is a continuous algebra homomorphism when $Hol(\Omega)$ is equipped with the topology of uniform convergence on compact subsets of Ω and $\mathcal{L}(\mathcal{H})$ is equipped with the operator norm. The map $f \mapsto f(T)$ is referred to as the Riesz Functional Calculus and there is a very clear treatment of it in [**Con90**].

We begin this section with some simple reformulations of the spectral set and rational dilation conditions.

LEMMA 1.6.2. *If $T \in \mathcal{L}(\mathcal{H})$ and $\sigma(T) \subset \Omega$, then the following conditions on T are equivalent.*

(i) $T \in \mathcal{S}$.
(ii) $Re\, h(T) \geq 0$ whenever $h \in \mathcal{R}at(\Omega^-)$ and $Re\, h \geq 0$ on Ω.
(iii) $Re\, h(T) \geq 0$ whenever $h \in \mathcal{C}$.

PROOF. The proof of Lemma 1.6.2 is based on two simple observations.

$$(1.6.3) \qquad z \mapsto \frac{1+z}{1-z} \text{ is a conformed map from}$$
$$\mathbb{D} \text{ onto the right half plane.}$$

$$(1.6.4) \qquad \text{If } S \in \mathcal{L}(\mathcal{H}) \text{ and } 1 \notin \sigma(S), \text{ then}$$
$$\|S\| \leq 1 \text{ if and only if } Re\, \frac{1+S}{1-S} \geq 0.$$

(1.6.3) is well known and (1.6.4) which is almost equally well known follows immediately by noting that

$$Re\, \frac{1+S}{1-S} = \frac{1}{2}\left(\frac{1+S^*}{1-S^*} + \frac{1+S}{1-S}\right)$$
$$= \frac{1}{2}(1-S^*)^{-1}((1+S^*)(1-S) + (1-S^*)(1+S))(1-S)^{-1}$$
$$= (1-S^*)^{-1}(1-S^*S)(1-S)^{-1}.$$

We shall prove Lemma 1.6.2 by showing in turn that (i) $\Rightarrow$ (ii), (ii) $\Rightarrow$ (iii), and (iii) $\Rightarrow$ (i).

First assume (i) holds and fix $h \in \mathcal{R}at(\Omega^-)$ with $Re\, h \geq 0$ on Ω. If $h = 0$, then trivially, $Re\, h(T) \geq 0$. Hence we may assume $h \neq 0$. But this implies via (1.6.3) that there exists $f \in \mathcal{R}at(\Omega^-)$ with $\|f\|_\infty \leq 1$ and

$$h = \frac{1+f}{1-f}.$$

Now, since $h \neq 0$, $f \neq 1$ and hence by the maximum principle, $1 \notin \sigma(f(T))$. Also, since $T \in \mathcal{S}$, $\|f(T)\| \leq 1$. Hence (1.5.6) implies that

$$\operatorname{Re} h(T) = \operatorname{Re} \frac{1+f(T)}{1-f(T)} \geq 0$$

and we have shown that (ii) holds.

Now assume that (ii) holds and fix $h \in \mathcal{C}$. Since $\operatorname{Re} h \geq 0$, a simple argument using Runge's Theorem yields a sequence $\{h_n\}$ with $h_n \in \mathcal{R}at(\Omega^-)$ for all n, $\operatorname{Re} h_n \geq 0$ for all n, and $h_n \to h$ in $Hol(\Omega)$. Then the continuity of the Riesz Functional Calculus implies that $h_n(T) \to h(T)$ in operator norm and (ii) implies that $\operatorname{Re} h_n(T) \geq 0$ for all n. Hence $\operatorname{Re} h(T) \geq 0$ and we have shown that (iii) holds.

Finally assume that (iii) holds and let $f \in \mathcal{R}at(\Omega^-)$ satisfy $\|f\|_\infty \leq 1$. For $r \in (0,1)$ define h_r by

$$h_r = \operatorname{Re} \frac{1+rf}{1-rf}.$$

Evidently, $h_r \neq 0$ and $\operatorname{Re} h_r \geq 0$ on Ω. Hence there exist $s,t \in \mathbb{R}$ with $s > 0$ and such that $sh_r + it \in \mathcal{C}$. By (iii) it follows that $\operatorname{Re} h_r(T) \geq 0$. Hence by (1.5.6), $\|rf(T)\| \leq 1$. Letting $r \to 1$ yields that $\|f(T)\| \leq 1$ and we have shown that (i) holds. $\square$

LEMMA 1.6.5. *If $T \in \mathcal{L}(\mathcal{H})$ and $\sigma(T) \subset \Omega$, then $T \in \mathcal{D}$ if and only if there exists a Hilbert space $\mathcal{K} \supseteq \mathcal{H}$ and there exists a normal operator $N \in \mathcal{L}(\mathcal{K})$ such that $\sigma(N) \subset \partial\Omega$ and*

(i) $\qquad \operatorname{Re} h(T) = P_\mathcal{H} \operatorname{Re} h(N) \mid \mathcal{H} \qquad$ *for all $h \in \mathcal{C} \cap \mathcal{R}at(\Omega^-)$.*

PROOF. If $T \in \mathcal{D}$, then from the definition there exist $\mathcal{K}$ and N with $h(T) = P_\mathcal{H} h(N) \mid \mathcal{H}$ for all $h \in \mathcal{R}at(\Omega^-)$. But if $S \in \mathcal{L}(\mathcal{K})$, then $P_\mathcal{H} S^* \mid \mathcal{H} = (P_\mathcal{H} S \mid \mathcal{H})^*$. Hence (i) holds. Now assume that there exist $\mathcal{K}$ and N with $\mathcal{K} \supset \mathcal{H}$, $N \in \mathcal{L}(\mathcal{K})$ normal, $\sigma(N) \subset \partial\Omega$, and such that (i) holds.

Fix $h \in \mathcal{R}at(\Omega^-)$. Since h is continuous on Ω^- and Ω is compact, there exists $c \in \mathbb{R}$ such that $\operatorname{Re}(h+c) > 0$ on Ω. Hence there exist $s,t \in \mathbb{R}$ with $s > 0$ such that $s(h+c) + it \in \mathcal{C}$. Hence by (i),

(1.6.6) $\qquad \operatorname{Re}(s(h+c)+it)(T) = P_\mathcal{H} \operatorname{Re}(s(h+c)+it)(N) \mid \mathcal{H}.$

But, $\operatorname{Re}(s(h+c)+it)(T) = s(\operatorname{Re} h(T)+c)$, $\operatorname{Re}(s(h+c)+it)(N) = s(\operatorname{Re} h(N)+c)$, and $P_\mathcal{H} c \mid \mathcal{H} = c$. Hence since $s \neq 0$ (1.6.6) implies that

(1.6.7) $\qquad \operatorname{Re} h(T) = P_\mathcal{H} \operatorname{Re} h(N) \mid \mathcal{H}.$

Summarizing, we have shown that (1.6.7) holds for all $h \in \mathcal{R}at(\Omega^-)$.

We now show that $T \in \mathcal{D}$. This will follow from the definition of $\mathcal{D}$ if we can show that

(1.6.8) $\qquad h(T) = P_\mathcal{H} h(N) \mid \mathcal{H}$

whenever $h \in \mathcal{R}at(\Omega^-)$. Accordingly, fix $h \in \mathcal{R}at(\Omega^-)$. Thus, (1.6.7) holds. But $ih \in \mathcal{R}at(\Omega^-)$ as well. Hence from (1.6.7) we deduce that

(1.6.9) $\qquad \operatorname{Im} h(T) = P_\mathcal{H} \operatorname{Im} h(N) \mid \mathcal{H}.$

Combining (1.6.7) and (1.6.9) gives (1.6.8) and the proof of Lemma 1.6.5 is complete. $\square$

Now if $N \in \mathcal{L}(\mathcal{K})$ is a normal operator then by the spectral theorem, N has a resolution
$$N = \int z\, dE(z)$$
for a unique spectral measure E on $\sigma(N)$. Furthermore, if u is a bounded Borel function on N then $u(N)$ can be defined by the formula,

(1.6.10) $$u(N) = \int u(z)\, dE(z).$$

The map $u \mapsto u(N)$ is a continuous $*$-homomorphism from the C^* algebra of bounded Borel functions on $\sigma(N)$ into $\mathcal{L}(\mathcal{K})$ that extends the Riesz Functional Calculus. It follows that if N is normal and $f \in \mathcal{R}at(\sigma(N))$, then

(1.6.11) $$\operatorname{Re} f(N) = (\operatorname{Re} f)(N).$$

Observe that (1.6.11) has the following consequence. If $N \in \mathcal{L}(\mathcal{K})$ is normal, $\sigma(N) \subset \partial\Omega$ and $f(T) = P_\mathcal{H} f(N) \mid \mathcal{H}$ for all $f \in \mathcal{R}at(\Omega^-)$, then in fact,

(1.6.12) $$\operatorname{Re} f(T) = P_\mathcal{H}(\operatorname{Re} f)(N) \mid \mathcal{H} \qquad \text{whenever } f \in \mathcal{R}at(\Omega^-).$$

Since $(\operatorname{Re} f)(N)$ might make sense when $\operatorname{Re} f(N)$ does not (e.g. $\operatorname{Re} f$ is a bounded Borel function on $\partial\Omega$ but f is not) it seems reasonable that the formula (1.6.12) could be generalized. One such generalization is nailed down precisely in the following lemma.

LEMMA 1.6.13. *Assume that $N \in \mathcal{L}(\mathcal{K})$ is normal, $\sigma(N) \subset \partial\Omega$, $\mathcal{H} \subset \mathcal{K}$, $T \in \mathcal{L}(\mathcal{H})$, $\sigma(T) \subset \Omega$, and $f(T) = P_\mathcal{H} f(N) \mid \mathcal{H}$ for all $f \in \mathcal{R}at(\Omega^-)$. If $f \in Hol(\Omega)$ and $\operatorname{Re} f \in C_\mathbb{R}(\partial\Omega)$ (i.e. $\operatorname{Re} f = u^\wedge$ for some $u \in C_\mathbb{R}(\partial\Omega)$), then*
$$\operatorname{Re} f(T) = P_\mathcal{H}(\operatorname{Re} f)(N) \mid \mathcal{H}.$$

PROOF. The lemma follows by taking a limit of (1.6.12). Since Ω is finitely connected and smoothly bounded, $\operatorname{Re} \mathcal{R}at(\Omega^-)$ is uniformly dense in $C_\mathbb{R}^h(\partial\Omega)$. Choose $f_n \in \mathcal{R}at(\Omega^-)$ such that $\operatorname{Re} f_n \to \operatorname{Re} f$ uniformly on $\partial\Omega$. Thus, by the continuity of the functional calculus defined in (1.6.10),

(1.6.14) $$(\operatorname{Re} f_n)(N) \to (\operatorname{Re} f)(N)$$

in operator norm. Now, if in addition we fix $z_0 \in \Omega$ and we choose f_n such that $\operatorname{Im} f_n(z_0) = \operatorname{Im} f(z_0)$ for all n, then the continuity of the harmonic conjugation operator on $Harm_\mathbb{R}^h(\Omega)$ implies that $f_n \to f$ in $Hol(\Omega)$. Hence, by the continuity of the Riesz functional calculus implies that

(1.6.15) $$f_n(T) \to f(T).$$

Lemma 1.6.13 now follows by combining (1.6.15), (1.6.12), and (1.6.14). $\square$

We now turn to the first goal of this section: to indicate precisely how the first Herglotz representation gives rise to a concrete condition for an operator to have Ω^- as a spectral set.

THEOREM 1.6.16. *If $T \in \mathcal{L}(\mathcal{H})$ and $\sigma(T) \subset \Omega$, then $T \in \mathcal{S}$ if and only if*

(1.6.17) $$\operatorname{Re} \Gamma_\alpha(T) \geq 0 \qquad \text{for all } \alpha \in T_\Omega.$$

1.6. THE HERGLOTZ REPRESENTATIONS AND OPERATOR THEORY

PROOF. By Lemma 1.6.2, $T \in \mathcal{S}$ if and only if $\operatorname{Re} h(T) \geq 0$ for all $h \in C$. Since $\Gamma_\alpha \in C$ for all $\alpha \in T_\Omega$, if $T \in \mathcal{S}$, then (1.6.17) holds. Conversely, assume that (1.6.17) holds and fix $h \in C$. By Theorem 1.3.27 there exists $\mu \in Prob(T_\Omega)$ such that
$$h = \int \Gamma_\alpha \, d\mu(\alpha).$$
Furthermore, since $\alpha \mapsto \Gamma_\alpha$ is a continuous map from T_Ω into $Hol(\Omega)$ the fact that $\sigma(T) \subset \Omega$ implies that
$$h(T) = \int \Gamma_\alpha(T) \, d\mu.$$
Hence, since (1.6.17) holds,
$$\operatorname{Re} h(T) = \int \operatorname{Re} \Gamma_\alpha(T) \, du \geq 0,$$
and the proof of Theorem 1.6.16 is complete. □

We now turn to the second goal of this section: to establish that the second Herglotz representation gives rise to a concrete condition for an operator to dilate to $\partial \Omega$.

THEOREM 1.6.18. *If $T \in \mathcal{L}(\mathcal{H})$ and $\sigma(T) \subset \Omega$, then $T \in \mathcal{D}$ if and only if there exist self-adjoint $\Phi_1, \ldots, \Phi_m \in \mathcal{L}(\mathcal{H})$ such that*

$$(1.6.19) \qquad \operatorname{Re} H_\lambda(T) + \sum_{r=1}^m \phi_r(\lambda) \Phi_r \geq 0 \qquad \text{for all } \lambda \in \partial \Omega.$$

PROOF. First suppose that $T \in \mathcal{D}$. Thus, there exists $\mathcal{K} \supset \mathcal{H}$ and a normal operator $N \in \mathcal{L}(\mathcal{K})$ such that $\sigma(N) \subset \partial \Omega$ and $f(T) = P_\mathcal{H} f(N) \mid \mathcal{H}$ for all $f \in \mathcal{R}at(\Omega^-)$. For each r define $\Phi_r \in \mathcal{L}(\mathcal{H})$ by

$$(1.6.20) \qquad \Phi_r = \phi_r(N).$$

We claim that (1.6.19) holds. To see this fix $\lambda \in \partial \Omega$ and let $\{w_\ell\}$ be a sequence in $C_\mathbb{R}(\partial \Omega)$ satisfying

$$(1.6.21) \qquad w_\ell \geq 0 \qquad \text{for each } \ell,$$

and

$$(1.6.22) \qquad w_\ell \sigma \to \delta_\lambda \qquad \text{in the } wk* \text{ topology}.$$

Set

$$(1.6.23) \qquad u_\ell = w_\ell - \sum_{r=1}^m \left(\int \phi_r w_\ell \, d\sigma \right) \phi_r \qquad \text{for each } \ell.$$

Notice that $\int u_\ell \phi_r \, d\sigma = 0$ for each r. Hence (1.2.10) implies that $u_\ell \in C_\mathbb{R}^h(\partial \Omega)$ for each ℓ. Choose $f_\ell \in Hol(\Omega)$ so that $\operatorname{Re} f_\ell = u_\ell^\wedge$ and $\tau(\operatorname{Im} f_\ell) = 0$.

Now, recalling (1.4.1), we see that (1.6.22) and (1.6.23) imply that $\mu_\ell \sigma \to \nu_\lambda$ in the $wk*$ topology. Hence, recalling Definition 1.4.2, we see that (1.2.5) and Lemma 1.2.20 imply that

$$(1.6.24) \qquad f_\ell \to H_\lambda \qquad \text{in } Hol(\Omega).$$

We now are able to establish that (1.6.19) holds. Note that

$$\operatorname{Re} H_\lambda(T) = \lim_\ell \operatorname{Re} f_\ell(T) \qquad \text{(by (1.6.24) and the continuity of the Riesz calculus)}$$

$$= \lim_\ell P_\mathcal{H} u_\ell(N) \mid \mathcal{H} \qquad \text{(by Lemma 1.6.13)}$$

$$= \lim_\ell P_\mathcal{H} \omega_\ell(N) \mid \mathcal{H} - \sum_{r=1}^m \phi_r(\lambda) P_\mathcal{H} \phi_r(N) \mid \mathcal{H} \qquad \text{(by (1.6.23) and (1.6.22))}$$

$$= \lim_\ell P_\mathcal{H} \omega_\ell(N) \mid \mathcal{H} - \sum_{r=1}^m \phi_r(\lambda) \Phi_r \qquad \text{(by (1.6.20))}$$

Also, (1.6.21) implies that $\omega_\ell(N) \geq 0$. Hence

$$\operatorname{Re} H_\lambda(T) + \sum_{r=1}^m \phi_r(\lambda) \Phi_r = \lim_\ell P_\mathcal{H} \omega_\ell(N) \mid \mathcal{H} \geq 0$$

and we have shown that (1.6.19) holds.

Now assume that (1.6.19) holds. For each Borel set $\Delta \subset \partial\Omega$ define $A(\Delta) \in \mathcal{L}(\mathcal{H})$ by the formula

$$(1.6.25) \qquad A(\Delta) = \int_\Delta \left(\operatorname{Re} H_\lambda(T) + \sum_{r=1}^m \phi_r(\lambda) \Phi_r \right) d\sigma(\lambda).$$

Evidently, (1.6.19) guarantees that $A(\Delta) \geq 0$ for each Borel set Δ. Also, the facts that $\int \phi_r \, d\sigma = 0$ for each r and $\operatorname{Re} \int H_\lambda \, d\sigma = \sigma^\wedge = 1$ imply that $A(\partial\Omega) = 1$. Hence A is an operator measure. It follows from the theorem of Naimark [**Nai43**] (or see [**Fil70**]) that there exist a Hilbert space $\mathcal{K} \supset \mathcal{H}$ and a $\mathcal{L}(\mathcal{K})$-valued spectral measure on the Borel subsets of $\partial\Omega$ such that $A(\Delta) = P_\mathcal{H} E(\Delta) \mid \mathcal{H}$ for all Borel sets $\Delta \subset \partial\Omega$. In particular, if $u \in C_\mathbb{R}(\partial\Omega)$, then

$$(1.6.26) \qquad \int u(\lambda) \, dA(\lambda) = P_\mathcal{H} \left(\int u(\lambda) \, dE(\lambda) \right) \mid \mathcal{H}.$$

Also, if we define $N \in \mathcal{L}(\mathcal{K})$ by the formula,

$$N = \int \lambda \, dE(\lambda),$$

then N is normal, $\sigma(N) \subset \partial\Omega$, and (1.6.10) holds for all $u \in C_\mathbb{R}(\partial\Omega)$.

We claim that $T \in \mathcal{D}$. This will follow from Lemma 1.6.5 if we can show that condition (i) of that lemma holds. Accordingly, fix $h \in C \cap \mathcal{R}at(\Omega^-)$. By Theorem 1.4.5, if we let $u = \operatorname{Re} h \mid \partial\Omega$,

$$h(z) = \int H_\lambda(z) u(\lambda) \, d\sigma(\lambda).$$

Hence, since $\lambda \mapsto H_\lambda$ is continuous from $\partial\Omega$ into $Hol(\Omega)$ and $\sigma(T) \subset \Omega$, we have that

$$h(T) = \int H_\lambda(T) u(\lambda) \, d\sigma(\lambda).$$

Note also that since $u = \operatorname{Re} h \mid \partial\Omega \in C_\mathbb{R}^h(\partial\Omega)$,

$$\int \phi_r u \, d\sigma = 0 \qquad \text{for each } r.$$

We now show that condition (i) of Lemma 1.6.5 holds for h.

$$\begin{aligned}
Re\, h(T) &= Re \int H_\lambda(T) u(\lambda)\, d\sigma \\
&= \int Re\, H_\lambda(T) u(\lambda)\, d\sigma \\
&= \int \left(Re\, H_\lambda(T) + \sum \phi_r(\lambda) \Phi_r \right) u(\lambda)\, d\sigma \\
&= \int u(\lambda)\, dA \\
&= P_\mathcal{H} \int u(\lambda)\, dE \mid \mathcal{H} \\
&= P_\mathcal{H} u(N) \mid \mathcal{H} \\
&= P_\mathcal{H} Re\, h(N) \mid \mathcal{H}.
\end{aligned}$$

This concludes the proof of Theorem 1.6.18. □

1.7. An Application

In this section we shall apply the results of previous sections to develop a computationally effective method for probing an open question in operator theory, namely Problem 1.1.4 from the introduction where $K = \Omega^-$, the closure of a bounded multiply-connected domain with analytic boundary. This method applies exclusively to the class of finite-dimensional operators which are diagonalizable and have one dimensional eigenspaces. Although the authors have proven that there is no loss of generality in restricting to this class—that is, the answer to Problem 1.1.4 is affirmative for all operators if and only if it is affirmative for operators in this class—this finding now seems moot in light of the experimental evidence of a finite-dimensional counterexample presented in Chapter 2 and [**Rap02**]. Thus we shall herein eschew this theoretical consideration, and instead focus on the foundations of the computational techniques that enable us to search for such a counterexample.

Historically, the ideas developed in this section were the basis for the authors' first systematic computational search for a counterexample. This search, however, produced only evidence in favor of an affirmative answer to Question 1.1.4. On one hand, such evidence gave us confidence in the consistency of our computational machinery. On the other hand, the failure to obtain a counterexample by such means motivated us to develop the more elaborate and ultimately successful computational methodology presented in Chapter 2. Although the techniques of Chapter 2 are dependent on much of the material of this section, an independent, if somewhat terse, exposition of such material is provided in Chapter 2 for readers who wish to skip this section (and perhaps refer back to it for details, when desired). Alternatively, this section can be read as a primer for Chapter 2. In addition, some results of independent theoretical interest are presented here, such as Theorems 1.7.56 and 1.7.80.

Recall from the introduction that Problem 1.1.4 has been resolved in the affirmative when $K = \Omega^-$ and genus(Ω^-) ≤ 1. As it currently stands, the problem is unresolved for general operators when genus(Ω) ≥ 2, although in [**Agl90**], [**Mis84**], and [**Pau87**] it was established that the problem has an affirmative answer when restricted to operators or spaces of dimension ≤ 2. Thus, the special case of operators on spaces of dimension 3 and genus(Ω) = 2 delineates the current theoretical frontier at which the equivalence of the spectral set and rational dilation conditions is unknown.

Because of the length and intricacy of this section, it is divided into five subsections. Sections 1.7.1–1.7.4 refine the spectral set and rational dilation criteria for certain finite-dimensional operators, using Theorems 1.6.16 and 1.6.18 as points of departure. In Section 1.7.1 we introduce a model for the finite-dimensional operators of interest, and show that for the purposes of probing the equivalence of the spectral set and rational dilation conditions we may restrict our attention to operators in this model which are extremal with respect to the constraint of having Ω^- as a spectral set. The advantages of focusing on such extremal operators are as follows.

(1) When searching for a counterexample, one is more likely to succeed by considering only extremal candidates.
(2) There is a straightforward, deterministic method for manufacturing extremal candidates. This is the focus of Section 1.7.3.
(3) For extremals, one achieves a dramatic reduction of the rational dilation condition of Theorem 1.6.18. This is the focus of Section 1.7.4.

Section 1.7.2 exploits an algebraic relationship among the functions $z \mapsto \Gamma_\alpha(z)$ (i.e. the extreme points of C) to reduce the spectral set condition of Theorem 1.6.16 from an $m+1$-parameter system of inequalities to an m-parameter system, thereby reducing the complexity of computational methods based on this theorem. Indeed, this reduction is invoked, in Section 1.7.3, to simplify the method for computing extremals. The algebraic relationship among the Γ_α is again utilized in Section 1.7.4 to reduce the rational dilation criterion for extremals. In the special case for operators on spaces of dimension 3, this reduction results in a rational dilation criterion which entails merely checking whether certain functions that are built up algebraically from $\phi_1, \ldots, \phi_m$ and H_λ either vanish identically or have nonnegative values. Section 1.7.5 describes how the results of Sections 1.7.1–1.7.4 can be combined to form the basis of an algorithm for searching for a counterexample to Problem 1.1.4.

The following notation and conventions will be assumed throughout this section. We shall assume that $\frac{d\sigma}{ds}$ is positive and real-analytic; thus $\phi_1, \ldots, \phi_m$ are real-analytic, and $\operatorname{Re} H_\lambda(z)$ is real-analytic in λ. For all $n \geq 1$, we shall make the identification

$$\mathcal{L}(\mathbb{C}^n) = M_n(\mathbb{C})$$

When we have need to express an operator $T \in \mathcal{L}(\mathbb{C}^n)$ in a basis B other than the standard basis, we shall denote the resultant matrix by $[T]_B$ to emphasize its dependence on B. For $n \geq 1$ fix distinct points $z_1, \ldots, z_n \in \Omega$, and write

$$z = (z_1, \ldots, z_n).$$

Define sets of operators $\mathcal{S}_z^0(\Omega) \subseteq M_n(\mathbb{C})$ and $\mathcal{D}_z^0(\Omega) \subseteq M_n(\mathbb{C})$ by

$$\mathcal{S}_z^0(\Omega) = \{T \in M_n(\mathbb{C}) : \sigma(T) = \{z_1, \ldots, z_n\} \text{ and } \Omega^- \text{ is a spectral set for } T\}$$

and
$$\mathcal{D}_z^0(\Omega) = \{T \in M_n(\mathbb{C}) : \sigma(T) = \{z_1, \ldots, z_n\}$$
and T has a rational dilation $\partial\Omega\}$.

Sections 1.7.1–1.7.4 will address the problem of developing a concrete method for probing the following restricted version of Problem 1.1.4:
$$\text{Is } \mathcal{D}_z^0(\Omega) = \mathcal{S}_z^0(\Omega)?$$

1.7.1. A model for operators in $\mathcal{S}_z^0(\Omega)$ and $\mathcal{D}_z^0(\Omega)$ and reduction to operators in this model which are extremal with respect to the spectral set condition. In this section we shall develop a model for operators in $\mathcal{S}_z^0(\Omega)$ and $\mathcal{D}_z^0(\Omega)$ in terms of the grammians of their eigenvectors. This model will result in sets of grammians, $S_z(\Omega)$ and $D_z(\Omega)$, which parameterize $\mathcal{S}_z^0(\Omega)$ and $\mathcal{D}_z^0(\Omega)$, respectively. In addition to admitting simpler descriptions than $\mathcal{S}_z^0(\Omega)$ and $\mathcal{D}_z^0(\Omega)$, $S_z(\Omega)$ and $D_z(\Omega)$ will be shown in Proposition 1.7.7 to have the distinct advantages of being convex and compact. These properties will be exploited in Theorem 1.7.24 to establish that, for the purposes of probing the question $\mathcal{S}_z^0(\Omega) \stackrel{?}{=} \mathcal{D}_z^0(\Omega)$, it suffices to restrict our attention to those elements of $S_z(\Omega)$ which are extremal with respect to the spectral set condition.

We begin by developing the model, which is based on the following elementary facts from linear algebra. If $M \in M_n(\mathbb{C})$ is diagonalizable with eigenvectors $v_1, \ldots, v_n$ and corresponding eigenvalues $z_1, \ldots, z_n$ respectively, then M is unitarily equivalent to

(1.7.1) $$G^{\frac{1}{2}} D_z G^{-\frac{1}{2}},$$

where $G = [\langle v_j, v_i \rangle]_{i,j=1}^n$ is the grammian of $v_1, \ldots, v_n$ and

$$D_z = \begin{pmatrix} z_1 & & & 0 \\ & z_2 & & \\ & & \ddots & \\ 0 & & & z_n \end{pmatrix}.$$

Note that G is not unique, since $v_1, \ldots, v_n$ may be scaled by arbitrary nonzero constants $c_1, \ldots, c_n$, respectively, resulting in a new grammian, G_1, which is Schur-equivalent to G—i.e.,

(1.7.2) $$G_1 = [\bar{c}_i c_j G_{ij}]_{i,j=1}^n$$
$$= D_c^* G D_c.$$

Moreover, if $z_1, \ldots, z_n$ are distinct and P and P_1 are positive matrices then $P^{\frac{1}{2}} D_z P^{-\frac{1}{2}}$ and $P_1^{\frac{1}{2}} D_z P_1^{-\frac{1}{2}}$ are unitarily equivalent if and only if P and P_1 are Schur-equivalent. Since for the purposes of probing the $\mathcal{D}_z^0(\Omega) \stackrel{?}{=} \mathcal{S}_z^0(\Omega)$ question it suffices to consider one member from each unitary equivalence class of $M_n(\mathbb{C})$, using the model (1.7.1) it suffices to consider one member from each Schur-equivalence class of positive matrices. Generically one way to single out such a member is to demand that it have entries equal to one on the main diagonal and nonnegative on the first row. This can be seen from (1.7.2), for if $G > 0$ with $G_{1j} \neq 0$ for all $j = 1, \ldots, n$ (which occurs generically) then there exists a vector $(c_1, \ldots, c_n) \in \mathbb{C}^n$ which is unique up to a unimodular scaling, such that $G_1 = D_c^* G D_c$ satisfies $(G_1)_{ii} = 1$ for all $i = 1, \ldots, n$, and $(G_1)_{1j} \geq 0$ for all j. Accordingly, we introduce the following definition.

DEFINITION 1.7.3. A (strictly) positive matrix $G \in M_n(\mathbb{C})$ is *normalized* if $G_{ii} = 1$ for all $i = 1, \ldots, n$ and $G_{1j} \geq 0$ for all $j = 2, \ldots, n$.

Note that the normalization introduced above differs slightly from the concept of "strong" normalization introduced in Section 2.2.3 of Chapter 2, which drops the condition $G_{1j} \geq 0$. There is no substantive difference between the two; the former results in the uniqueness assertion of Proposition 1.7.4 below, while the latter is simpler for purposes of numerical computation.

The ideas thus far discussed in Section 1.7.1 are summarized in the following result.

PROPOSITION 1.7.4. Let $z = (z_1, \ldots, z_n)$ with $z_1, \ldots, z_n \in \mathbb{C}$ distinct and let $T \in M_n(\mathbb{C})$ with $\sigma(T) = \{z_1, \ldots, z_n\}$. There exists a normalized $G > 0$ such that T is unitarily equivalent to $G^{\frac{1}{2}} D_z G^{-\frac{1}{2}}$. Generically, G is unique. Further, G is equal to the grammian of the eigenvectors $v_1, \ldots, v_n$ of T corresponding to $z_1, \ldots, z_n$, normalized so that $\|v_i\| = 1$ for all $i = 1, \ldots n$ and $\langle v_j, v_1 \rangle \geq 0$ for all $j = 2, \ldots, n$.

Fix a domain $\Omega \subseteq \mathbb{C}$ of genus m. Let $n \geq 3$ and let $z = (z_1, \ldots, z_n)$ where $z_1, \ldots, z_n \in \Omega$ are distinct. Define sets of normalized grammians $S_z(\Omega)$ and $D_z(\Omega)$ by

$$S_z(\Omega) = \{G > 0 : G \text{ is normalized and } G^{\frac{1}{2}} D_z G^{-\frac{1}{2}} \in \mathcal{S}_z^0(\Omega)\}$$

and

$$D_z(\Omega) = \{G > 0 : G \text{ is normalized and } G^{\frac{1}{2}} D_z G^{-\frac{1}{2}} \in \mathcal{D}_z^0(\Omega)\}.$$

Clearly, by Proposition 1.7.4 and by the unitary invariance of $\mathcal{S}_z^0(\Omega)$ and $\mathcal{D}_z^0(\Omega)$, $\mathcal{S}_z^0(\Omega) = \mathcal{D}_z^0(\Omega)$ if and only if $S_z(\Omega) = D_z(\Omega)$. Hereafter in Section 1.7 we shall focus on the sets $S_z(\Omega)$ and $D_z(\Omega)$, rather than $\mathcal{S}_z^0(\Omega)$ and $\mathcal{D}_z^0(\Omega)$. The advantages of $S_z(\Omega)$ and $D_z(\Omega)$ will be revealed in Proposition 1.7.7 below. For the purposes of facilitating the statement of this proposition, we introduce the following notation.

If $f \in Hol(\Omega)$ and $z = (z_1, \ldots, z_n)$ we shall write

$$f(z) = (f(z_1), \ldots, f(z_n)),$$
$$\overline{f(z)} = (\overline{f(z_1)}, \ldots, \overline{f(z_n)}).$$

For $\alpha \in T_\Omega$, define an $n \times n$ matrix $\gamma(\alpha) = [\gamma_{ij}(\alpha)]_{i,j=1}^n$ by setting

(1.7.5) $$\gamma_{ij}(\alpha) = \frac{1}{2}\left(\overline{\Gamma_\alpha(z_i)} + \Gamma_\alpha(z_j)\right).$$

Similarly, for $\lambda \in \partial\Omega$, define $h(\lambda) = [h_{ij}(\lambda)]_{i,j=1}^n$ by

(1.7.6) $$h_{ij}(\lambda) = \frac{1}{2}\left(\overline{H_\lambda(z_i)} + H_\lambda(z_j)\right).$$

If $M, N \in M_n(\mathbb{C})$ let $M * N$ denote the matrix whose (i,j) entry is given by

$$(M * N)_{ij} = M_{ij} N_{ij},$$

the Schur product of M and N.

PROPOSITION 1.7.7.

(1) $G \in S_z(\Omega)$ if and only if G is a normalized positive matrix and $\gamma(\alpha) * G \geq 0$ for all $\alpha \in T_\Omega$.

(2) $G \in D_z(\Omega)$ if and only if G is a normalized *positive* matrix and there exists $\Psi \in (M_n(\mathbb{C}))^m$ with $\Psi_1, \ldots, \Psi_m$ self-adjoint, such that

(1.7.8) $$h(\lambda) * G + \sum_{r=1}^{m} \varphi_r(\lambda)\Psi_r \geq 0 \quad \text{for all } \lambda \in \partial\Omega.$$

(3) $S_z(\Omega)$ and $D_z(\Omega)$ are convex, compact subsets of $M_n(\mathbb{C})$.

The proof of Proposition 1.7.7 will be based on the following five lemmas. The reader can verify the first result by direct computation.

LEMMA 1.7.9. *Let $G > 0$, $f \in Hol(\Omega)$, and define the $n \times n$ matrix F by*

(1.7.10) $$F = \left[\frac{1}{2}(\overline{f(z_i)} + f(z_j))\right]_{i,j=1}^{n}.$$

The following formula obtains:

(1.7.11) $$\operatorname{Re} f(G^{\frac{1}{2}} D_z G^{-\frac{1}{2}}) = G^{-\frac{1}{2}}(F * G)G^{-\frac{1}{2}}.$$

In particular, $\operatorname{Re} f(G^{\frac{1}{2}} D_z G^{-\frac{1}{2}}) \geq 0$ if and only if $F * G \geq 0$.

LEMMA 1.7.12. *If $G \in M_n(\mathbb{C})$ is a normalized positive matrix then $\|G\| < n$.*

PROOF. Assuming $G > 0$ and $G_{ii} = 1$ for all $i = 1, \ldots, n$,

$$\begin{aligned}
\|G\| &= \max \text{eigenvalue}(G) \\
&< \text{sum (eigenvalues}(G)) \\
&= \operatorname{Tr}(G) \\
&= n.
\end{aligned}$$

$\square$

LEMMA 1.7.13. *For every $z = (z_1, \ldots, z_n)$ with $z_1, \ldots, z_n \in \Omega$ distinct there exists a positive constant c_z, which depends only on z and Ω, such that*

$$G \geq c_z I$$

for all $G \in S_z(\Omega)$.

PROOF. We shall argue by contradiction. Accordingly, assume there exists $z = (z_1, \ldots, z_n)$ with $z_1, \ldots, z_n \in \Omega$ distinct and there exist $\{G_j\} \subseteq S_z(\Omega)$ and $\{w_j\} \subseteq \mathbb{C}^n$ such that

$$\|w_j\| = 1 \quad \text{for all } j = 1, 2, \ldots$$

and

(1.7.14) $$G_j w_j \to 0.$$

By compactness of the unit sphere of $\mathbb{C}^n$, we may assume, by passing to a subsequence of $\{w_j\}$ if necessary, that

$$w_j \to v \in \mathbb{C}^n \quad \text{with } \|v\| = 1.$$

For some $n_0 \in \{1, \ldots, n\}$ we must have $|v_{n_0}| \geq 1/\sqrt{n}$; without loss of generality we may assume

$$|v_1| \geq 1/\sqrt{n}.$$

Thus there exists $J \geq 1$ such that

(1.7.15) $$|(w_j)_1| \geq \frac{1}{2\sqrt{n}} \quad \text{for all } j \geq J.$$

Since $z_1, \ldots, z_n$ are distinct there exists a polynomial p such that
$$p(z_i) = \begin{cases} 1 & \text{if } i = 1 \\ 0 & \text{if } i = 2, \ldots, n. \end{cases}$$

Define
$$T_j = G_j^{\frac{1}{2}} D_z G_j^{-\frac{1}{2}} \quad j = 1, 2, \ldots,$$
$$u_j = G_j^{\frac{1}{2}} w_j \quad j = 1, 2, \ldots,$$

and
$$e_1 = \begin{pmatrix} 1 \\ 0 \\ \vdots \\ 0 \end{pmatrix}.$$

We have
$$p(T_j) u_j = G_j^{\frac{1}{2}} D_{p(z)} G_j^{-\frac{1}{2}} (G_j^{\frac{1}{2}} w_j)$$
$$= G_j^{\frac{1}{2}} \begin{pmatrix} 1 & & & 0 \\ & 0 & & \\ & & \ddots & \\ 0 & & & 0 \end{pmatrix} w_j$$
$$= (w_j)_1 G_j^{\frac{1}{2}} e_1.$$

Hence if $j \geq J$, then

(1.7.16) $$\|p(T_j) u_j\|^2 = |(w_j)_1|^2 \langle G_j^{\frac{1}{2}} e_1, G_j^{\frac{1}{2}} e_1 \rangle$$
$$= |(w_j)_1|^2 \langle G_j e_1, e_1 \rangle$$
$$= |(w_j)_1|^2 \quad \text{(since } (G_j)_{11} = 1\text{)}$$
$$\geq \frac{1}{4n} \quad \text{(by (1.7.15))}.$$

Now by (1.7.14), $0 = \lim_{j \to \infty} \langle G_j w_j, w_j \rangle = \lim_{j \to \infty} \langle G_j^{\frac{1}{2}} w_j, G_j^{\frac{1}{2}} w_j \rangle = \lim_{j \to \infty} \|u_j\|^2$; thus $u_j \to 0$. Consequently, (1.7.16) implies
$$\|p(T_j)\| \to \infty.$$

But since $G_j \in S_z(\Omega)$, Ω^- is a spectral set for T_j, so that for all j $\|p(T_j)\| \leq \sup_{z \in \Omega}(p(z)) < \infty$. This contradiction completes the proof of Lemma 1.7.13. $\square$

LEMMA 1.7.17. *(Properties of $h(\lambda) * G + \sum_{r=1}^m \varphi_r(\lambda) \Psi_r$).* Let $\Psi \in (M_n(\mathbb{C}))^m$.
(1) For all $\alpha \in T_\Omega$

(1.7.18) $$\sum_{s=0}^m w_s(\alpha) \left(h(\lambda) * G + \sum_{r=1}^m \varphi_r(\alpha_s) \Psi_r \right) = \gamma(\alpha) * G$$

where $w_0(\alpha), \ldots, w_m(\alpha)$ are given by (1.3.21).

(2) For all $r \in \{1, \ldots, m\}$

(1.7.19) $$\Psi_r = \int_{\partial\Omega} \phi_r(\lambda)\left(h(\lambda) * G + \sum_{r=1}^{m} \varphi_r(\lambda)\Psi_r\right) d\sigma(\lambda).$$

(3) If $(h(\lambda) * G + \sum_{r=1}^{m} \varphi_r(\lambda)\Psi_r)$ is self-adjoint for all $\lambda \in \partial\Omega$, then $\Psi_1, \ldots, \Psi_m$ are self-adjoint. In particular, if $(h(\lambda) * G + \sum_{r=1}^{m} \varphi_r(\lambda)\Psi_r) \geq 0$ for all λ, then $\Psi_1, \ldots, \Psi_m$ are self-adjoint.

PROOF. (1) follows from the identity $\Gamma_\alpha = \sum_{s=0}^{m} w_s(\alpha) H_{\alpha_s}$ (formula (1.4.9)), and Lemma 1.3.20, which implies that for all $\alpha \in T_\Omega$ and all $r = 1, \ldots, m$, $\sum_{s=0}^{m} w_s(\alpha)\phi_r(\alpha_s) = 0$.

To prove (2) note that

$$\int_{\partial\Omega} \phi_r(\lambda)\left(h(\lambda) * G + \sum_{s=1}^{m} \varphi_s(\lambda)\Psi_s\right) d\sigma(\lambda)$$
$$= \left[\frac{1}{2}G_{ij} \int_{\partial\Omega} (\overline{H_\lambda(z_i)} + H_\lambda(z_j))\phi_r(\lambda) d\sigma(\lambda)\right]_{i,j=1}^{n} + \sum_{s=1}^{m} \Psi_s \int_{\partial\Omega} \phi_s\phi_r \, d\sigma.$$

Formula (1.7.19) now follows by invoking the orthonormality of $\phi_1, \ldots, \phi_m$ and observing that since

$$\operatorname{Re}\int_{\partial\Omega} H_\lambda(z)\phi_r(\lambda) d\sigma(\lambda) = \int_{\partial\Omega} (P_z^\sigma(\lambda) - \sum_{s=1}^{m} \phi_s(\lambda)\phi_s^\wedge(z))\phi_r(\lambda) d\sigma(\lambda)$$
$$= \phi_r^\wedge(z) - \phi_r^\wedge(z) = 0,$$

the holomorphic function $z \mapsto \int_{\partial\Omega} H_\lambda(z)\phi_r(\lambda) d\sigma(\lambda)$ must be identically equal to an imaginary constant, implying the first matrix on the right hand side of (1.7.20) must vanish.

(3) follows from (2) and the fact that the self-adjoint matrices form a real subspace of $M_n(\mathbb{C})$. This completes the proof of Lemma 1.7.17. □

Define

$$P_z(\Omega) = \{\Psi \in (M_n(\mathbb{C}))^m : \Psi_1, \ldots, \Psi_m \text{ are self-adjoint and there exists}$$
$$G \in D_z(\Omega) \text{ such that for every } \lambda \in \partial\Omega \ h(\lambda) * G + \sum_{r=1}^{m} \varphi_r(\lambda)\Psi_r \geq 0\}.$$

For $\Psi \in (M_n(\mathbb{C}))^m$ define

$$\|\Psi\| = \max\{\|\Psi_r\| : r = 1, \ldots, m\}.$$

LEMMA 1.7.20. For every $z = (z_1, \ldots, z_n)$ with $z_1, \ldots, z_n \in \Omega$ distinct, there exists a positive constant d_z, which depends only on z and Ω, such that

$$\|\Psi\| \leq d_z \quad \text{for all } \Psi \in P_z(\Omega).$$

PROOF. Fix distinct $z_1, \ldots, z_n \in \Omega$, and fix $\Psi \in P_z(\Omega)$. Let $G \in D_z(\Omega)$ satisfy $h(\lambda) * G + \sum_{r=1}^{m} \varphi_r(\lambda)\Psi_r \geq 0$ for all $\lambda \in \partial\Omega$. By formula (1.7.18) and positivity of the weights $w_0, \ldots, w_m$ we have, for all $s = 0, \ldots, m$ and all $\alpha \in T_\Omega$,

$$\left\| w_s(\alpha) \left(h(\alpha_s) * G + \sum_{r=1}^{m} \varphi_r(\alpha_s)\Psi_r \right) \right\| \leq \|\gamma(\alpha) * G\|.$$

Defining
$$B = \sup\{|\Gamma_\alpha(z_j)| : \alpha \in T_\Omega, \quad j = 1, \ldots, n\}$$
and
$$b = \inf\{w_r(\alpha) : \alpha \in T_\Omega, \quad r = 0, \ldots, m\}$$
we have $B < \infty$, by continuity of $\alpha \mapsto \Gamma_\alpha(z)$, and $b > 0$, by positivity and continuity of $\alpha \mapsto w_r(\alpha)$. Since $G > 0$ and $G_{ii} = 1$ for all i, $|G_{ij}| \leq 1$ for all i, j. Therefore, for all $\lambda \in \partial\Omega$,

$$\left\| h(\lambda) * G + \sum_{r=1}^{m} \varphi_r(\lambda)\Psi_r \right\| \leq \frac{1}{b} \sup_{\alpha \in T_\Omega} \|\gamma(\alpha) * G\|$$

$$\leq \frac{1}{b} \sup_{\alpha \in T_\Omega} \left(\sum_{i,j=1}^{n} \left| \frac{1}{2} G_{ij} (\overline{\Gamma_\alpha(z_i)} + \Gamma_\alpha(z_j)) \right|^2 \right)^{\frac{1}{2}}$$

$$\leq \frac{nB}{b}.$$

Formula (1.7.19) hence yields, for all $r = 1, \ldots, m$,

$$\|\Psi_r\| \leq \int_{\partial\Omega} |\phi_r(\lambda)| \left\| h(\lambda) * G + \sum_{r=1}^{m} \varphi_r(\lambda)\Psi_r \right\| d\sigma(\lambda)$$

$$\leq \frac{nB}{b} \|\phi_r\|_{L^2_{\mathbb{R}}(\sigma)} = \frac{nB}{b}.$$

The proof of Lemma 1.7.20 is completed by setting $d_z = \frac{nB}{b}$ and noting that B depends only on z and Ω, while b depends only on Ω.

We are now ready to prove Proposition 1.7.7.

(1) follows from Theorem 1.6.16 and Lemma 1.7.9.

(2) follows from Theorem 1.6.18 and Lemma 1.7.9 upon relating $\Psi_1, \ldots, \Psi_m$ and $\Phi_1, \ldots, \Phi_m$ by means of the equations

$$\Psi_r = G^{\frac{1}{2}} \Phi_r G^{-\frac{1}{2}} \qquad r = 1, \ldots, m.$$

To prove the convexity assertions of part (3) of Proposition 1.7.7, first note that the set of normalized positive matrices is convex. The convexity of $S_z(\Omega)$ now follows from part (1) of Proposition 1.7.7 and the linearity of $G \mapsto \gamma(\alpha) * G$. Likewise, the convexity of $D_z(\Omega)$ follows form part (2) and the linearity of $(G, \Psi) \mapsto h(\lambda) * G + \sum_{r=1}^{m} \varphi_r(\lambda)\Psi_r$.

The reader can verify the compactness of $S_z(\Omega)$ by invoking the Heine–Borel Theorem, together with part (1) of Proposition 1.7.7, Lemmas 1.7.12 and 1.7.13, and the continuity of $G \mapsto \gamma(\alpha) * G$. Similarly, the compactness of $D_z(\Omega)$ may be deduced from the inclusion $D_z(\Omega) \subseteq S_z(\Omega)$ together with part (2) of Proposition 1.7.7, Lemma 1.7.20, and the continuity of $(G, \Psi) \mapsto h(\lambda) * G + \sum_{r=1}^{m} \varphi_r(\lambda)\Psi_r$. This completes the proof of Proposition 1.7.7. □

We remark that by part (3) of Lemma 1.7.17, the requirement that $\Psi_1, \ldots, \Psi_m$ be self-adjoint in the statement of Proposition 1.7.7 is superfluous. However we have included explicit mention of self-adjointness to emphasize that when developing a computation technique which attempts to find $\Psi_1, \ldots, \Psi_m$ satisfying (1.7.8) one may restrict the search exclusively to self-adjoint candidates.

We conclude Section 1.7.1 by introducing an extremal problem on $S_z(\Omega)$ and showing, in Theorem 1.7.24 below that when probing the question $\mathcal{S}_z^0(\Omega) \stackrel{?}{=} \mathcal{D}_z^0(\Omega)$ it suffices to restrict our attention to those elements of $S_z(\Omega)$ which solve this extremal problem. In Chapter 2, such grammians (with a slightly weaker normalization) will be referred to as extremals.

Let $Re\, M_n(\mathbb{C})$ denote the self-adjoint $n \times n$ matrices over $\mathbb{C}$. Define $g_z : Re\, M_n(\mathbb{C}) \to \mathbb{R}$ by

$$(1.7.21) \qquad g_z(A) = \inf_{\alpha \in T_\Omega} \lambda_{\min}(A * \gamma(\alpha)),$$

where $\lambda_{\min}(B)$ denotes the minimum eigenvalue of a self-adjoint matrix B. By Proposition 1.7.7,

$$S_z(\Omega) = \{G > 0 : G \text{ is normalized and } g_z(G) \geq 0\}.$$

Define

$$(1.7.22) \qquad \partial S_z(\Omega) = \{G \in S_z(\Omega) : g_z(G) = 0\},$$

and let $ext\, S_z(\Omega)$ denote the set of extreme points of $S_z(\Omega)$.

LEMMA 1.7.23. *If $n \geq 3$ then $ext\, S_z(\Omega) \subseteq \partial S_z(\Omega)$.*

PROOF. We shall argue that if $G \notin \partial S_z(\Omega)$ then $G \notin ext\, S_z(\Omega)$. Accordingly, let $G \in S_z(\Omega) \setminus \partial S_z(\Omega)$, so that $g_z(G) > 0$. Let $\Delta \in M_n(\mathbb{C})$ be any nonzero self-adjoint matrix which vanishes along the principal diagonal and the first row (note that $n \geq 3$ is essential here). Since $G > 0$ and g_z is continuous, there exists $t > 0$ such that $G \pm t\Delta$ is a normalized positive matrix and $g_z(G \pm t\Delta) > 0$; consequently, $G \pm t\Delta \in S_z(\Omega)$. But $G = \frac{1}{2}(G + t\Delta) + \frac{1}{2}(G - t\Delta)$, thus $G \notin ext\, S_z(\Omega)$. □

THEOREM 1.7.24. *Let $\Omega \subseteq \mathbb{C}$ be a multiply-connected domain, $n \geq 3$, and $z = (z_1, \ldots, z_n)$ with $z_1, \ldots, z_n \in \Omega$ distinct. $\mathcal{S}_z^0(\Omega) = \mathcal{D}_z^0(\Omega)$ if and only if $\partial S_z(\Omega) \subseteq D_z(\Omega)$.*

PROOF. Since, as already noted, $\mathcal{S}_z(\Omega) = \mathcal{D}_z(\Omega)$ if and only if $S_z(\Omega) = D_z(\Omega)$, and since $D_z(\Omega) \subseteq S_z(\Omega)$ a priori, we need only show that $\partial S_z(\Omega) \subseteq D_z(\Omega)$ implies $S_z(\Omega) \subseteq D_z(\Omega)$. Accordingly, suppose $\partial S_z(\Omega) \subseteq D_z(\Omega)$. By Lemma 1.7.23 we have

$$ext\, S_z(\Omega) \subseteq D_z(\Omega).$$

But by Proposition 1.7.7, $S_z(\Omega)$ is a convex, compact subset of the finite-dimensional space $M_n(\mathbb{C})$, and $D_z(\Omega)$ is convex, hence

$$S_z(\Omega) = cohull(ext\, S_z(\Omega)) \subseteq D_z(\Omega).$$

□

1.7.2. Reduction of the spectral set condition from a $m+1$-parameter system of inequalities to an m-parameter system of inequalities.

Proposition 1.7.7 suggests an effective computational method for testing whether a normalized positive matrix belongs to $S_z(\Omega)$, namely, checking the validity of a system of operator inequalities that is parameterized by the $m+1$ dimensional torus $T_\Omega = \partial_0 \times \cdots \times \partial_m$. In this section we shall show that, in fact, it suffices to check a proper subset of such inequalities that is parameterized by an m-dimensional subtorus of T_Ω. This is a significant reduction for the purposes of machine computation.

Specifically, define for each $\lambda_0 \in \partial\Omega$, the m-torus $T_\Omega(\lambda_0) \subseteq T_\Omega$ by

$$T_\Omega = \{\alpha \in T_\Omega : \alpha_r = \lambda_0\},$$

where r is the unique element of $\{0,\ldots,m\}$ such that $\lambda_0 \in \partial_r$. We shall establish the following reduction of part (1) of Proposition 1.7.7.

PROPOSITION 1.7.25. Fix $\lambda_0 \in \partial\Omega$. $G \in S_z(\Omega)$ if and only if G is a normalized positive matrix and

(1.7.26) $$\gamma(\alpha) * G \geq 0 \qquad \text{for all } \alpha \in T_\Omega(\lambda_0).$$

The remainder of this section is devoted to the proof of Proposition 1.7.25, which will make use of two preliminary results, Lemmas 1.7.28 and 1.7.29 below.

Let $\mathbb{R}^* = \mathbb{R} \cup \{\infty\}$ denote the one point compactification of the real numbers. By Proposition 1.3.31, for every $\alpha \in T_\Omega$ and every $t \in \mathbb{R}^*$, there exists a unique element of T_Ω, which we shall denote by $\tau_t(\alpha)$, such that

(1.7.27) $$\Gamma_\alpha((\tau_t(\alpha))_r) = it, \qquad r = 0, 1, \ldots, m.$$

where $(\tau_t(\alpha))_r$ denotes the rth component of $\tau_t(\alpha)$. Note that

$$\tau_\infty(\alpha) = \alpha,$$

and for each $\alpha \in T_\Omega$ and each $r \in \{0, 1, \ldots, m\}$,

$$t \mapsto (\tau_t(\alpha))_r$$

maps $\mathbb{R}^*$ homeomorphically onto ∂_r.

The following lemma describes a simple algebraic relationship between Γ_α and $\Gamma_{\tau_t(\alpha)}$. To express this relationship, we introduce the following notation: for $f, g \in Hol(\Omega)$ we shall write

$$f \simeq g$$

if there exists $a > 0$ and $b \in \mathbb{R}$ such that

$$f = ag + ib.$$

It is easy to see that $\simeq$ is an equivalence relation on $Hol(\Omega)$.

LEMMA 1.7.28. If $\alpha \in T_\Omega$ and $t \in \mathbb{R}$ then

$$\Gamma_{\tau_t(\alpha)} \simeq \frac{1}{\Gamma_\alpha - it}.$$

PROOF. Fix $\alpha \in T_\Omega$ and $t \in \mathbb{R}$. Note that the map $\xi \mapsto \frac{1}{\xi - it}$ is an automorphism of the right half plane. Thus since Γ_α satisfies (1.3.32) and (1.3.33), so does $\frac{1}{\Gamma_\alpha - it}$. By Proposition 1.3.31, there exists a unique $\beta \in T_\Omega$ such that $\frac{1}{\Gamma_\alpha - it} \simeq \Gamma_\beta$, and moreover, the components of β coincide with the poles of $\frac{1}{\Gamma_\alpha - it}$ on $\partial\Omega$. But the poles of $\frac{1}{\Gamma_\alpha - it}$ on $\partial\Omega$ are precisely the zeros of $\Gamma_\alpha - it$ on $\partial\Omega$, hence $\Gamma_\alpha(\beta_r) = it$ for all $r = 0, \ldots, m$. By (1.7.27) and the uniqueness of β we must have $\beta = \tau_t(\alpha)$. □

The operator-theoretic significance of the equivalence relation $\simeq$ is revealed by the following result.

LEMMA 1.7.29.
(1) For every $\alpha \in T_\Omega$ and every $t \in \mathbb{R}^*$ there exists $a > 0$ such that
$$\gamma(\tau_t) = a(U_t \otimes U_t) * \gamma(\alpha) \tag{1.7.30}$$
where $U_t = (\frac{1}{\Gamma_\alpha(z_1)-it}, \ldots, \frac{1}{\Gamma_\alpha(z_n)-it})$.
(2) If $G \in M_n(\mathbb{C})$, $\alpha \in T_\Omega$ and $t \in \mathbb{R}^*$, then $\gamma(\alpha) * G \geq 0$ if and only if $\gamma(\tau_t(\alpha)) * G \geq 0$.

PROOF. (1) Let $t \in \mathbb{R}^*$, $\alpha \in T_\Omega$. By Lemma 1.7.28 there exists $a > 0$, $b \in \mathbb{R}$ such that $\Gamma_{\tau_t(\alpha)} = \frac{a}{\Gamma_\alpha - it} + ib$. Hence

$$\gamma(\tau_t) = \frac{1}{2}\left[\frac{a}{\Gamma_\alpha(z_k) - it} + \frac{a}{\Gamma_\alpha(z_j) - it}\right]_{k,j=1}^n$$
$$= \frac{a}{2}\left[\frac{1}{\Gamma_\alpha(z_k) - it}\frac{1}{\Gamma_\alpha(z_j) - it}\right]_{k,j=1}^n$$
$$* \left[\overline{(\Gamma_\alpha(z_k) - it)} + \Gamma_\alpha(z_j) - it\right]_{k,j=1}^n$$
$$= a(U_t \otimes U_t) * \gamma(\alpha).$$

(2) follows from (1) and the fact that if $U \in \mathbb{C}^n$ has all nonzero entries and $M \in M_n(\mathbb{C})$ then $(U \otimes U) * M \geq 0$ if and only if $M \geq 0$.

It should be noted that the above lemma has an infinite-dimensional analog which references an operator $T \in \mathcal{L}(H)$ directly, rather than the grammian G. (1.7.30) is replaced by the formula

$$\operatorname{Re}\Gamma_{\tau_t(\alpha)}(T) = a\left(\frac{1}{\Gamma_\alpha(T) - it}\right)^* \operatorname{Re}\Gamma_\alpha(T) \frac{1}{\Gamma_\alpha(T) - it},$$

(where $a > 0$ depends on T and t) which, as the reader may easily verify, holds for all $T \in \mathcal{L}(H)$ with $\sigma(T) \subseteq \Omega$, all $\alpha \in T_\Omega$ and all $t \in \mathbb{R}$. Using this formula one easily obtains an analog of part (2) of Lemma 1.7.29:

if $\sigma(T) \subseteq \Omega$, $\alpha \in T_\Omega$ and $t \in \mathbb{R}^*$, then $\operatorname{Re}\Gamma_{\tau_t(\alpha)}(T) \geq 0$
if and only if $\operatorname{Re}\Gamma_\alpha(T) \geq 0$.

We are now ready to prove Proposition 1.7.25. Let $G > 0$ be a normalized positive $n \times n$ matrix, and let $\lambda_0 \in \partial\Omega$. By virtue of Proposition 1.7.7, it suffices to show that given any $\beta \in T_\Omega$, the inequalities (1.7.26) imply that $\gamma(\beta) * G \geq 0$. Accordingly, assume (1.7.26) and let $\beta \in T_\Omega$. Let r be the unique element of $\{0, \ldots, m\}$ such that $\lambda_0 \in \partial_r$ and let

$$t = \frac{1}{i}\Gamma_\beta(\lambda_0)$$

so that $t \in \mathbb{R}$ if $\lambda_0 \neq \beta_r$ and $t = \infty$ if $\lambda_0 = \beta_r$. Now

$$\Gamma_\beta(\lambda_0) = it$$

and by (1.7.27),
$$\Gamma_\beta\big((\tau_t(\beta))_r\big) = it.$$
By Proposition 1.3.31, $\lambda \mapsto \Gamma_\beta(\lambda)$ is 1-1 on ∂_r. Thus we must have $(\tau_t(\beta))_r = \lambda_0$, implying $\tau_t(\beta) \in T_\Omega(\lambda_0)$. The inequalities (1.7.26) together with part (2) of Lemma 1.7.29 imply $\gamma(\beta) * G \geq 0$, and the proof of Proposition 1.7.25 is complete. □

We remark that by the discussion immediately following Lemma 1.7.29, we can easily adapt the proof of Proposition 1.7.25 to obtain the following infinite-dimensional analog of Proposition 1.7.25, a strengthening of Theorem 1.6.16:

Fix $\lambda_0 \in \partial\Omega$. If $T \in \mathcal{L}(H)$, and $\sigma(T) \subseteq \Omega$, then

$T \in \mathcal{S}$ if and only if $Re\,\Gamma_\alpha(T) \geq 0$ for all $\alpha \in T_\Omega(\lambda_0)$.

1.7.3. Manufacturing grammians which are extremal with respect to the spectral set condition. In this section we shall evolve a method, described in Proposition 1.7.36 below, for producing elements in the set $\partial S_z(\Omega)$ defined by (1.7.22). In Theorem 1.7.42 it is shown that this method is general, in the sense that every element of $\partial S_z(\Omega)$ can be derived in this way; by Theorem 1.7.24, the question $\mathcal{S}_z^0(\Omega) \stackrel{?}{=} \mathcal{D}_z^0(\Omega)$ reduces to the question of whether each grammian arising from this construction belongs to $D_z(\Omega)$.

The following definition isolates two useful concepts for identifying when a matrix $G \in S_z(\Omega)$ belongs to $\partial S_z(\Omega)$.

DEFINITION 1.7.31. Let $G \in S_z(\Omega)$. We say that $\alpha \in T_\Omega$ is an *extremal point of* G if $\ker \gamma(\alpha) * G \neq (0)$. If α is an extremal point of G and u is a nonzero element of $\ker \gamma(\alpha) * G$, we say that u is an *extremal vector* of G, or, alternatively, that u is an *extremal vector of* G *at* α.

By continuity of the function $g_z : Re\,M_n(\mathbb{C}) \to \mathbb{R}$ defined by (1.7.21) we obtain the following result.

LEMMA 1.7.32. *If* $G \in S_z(\Omega)$ *then* $G \in \partial S_z(\Omega)$ *if and only if there exists* $\alpha \in T_\Omega$ *which is an extremal point of* G.

For the purposes of stating Proposition 1.7.36 we define $M_z : Re\,M_n(\mathbb{C}) \times T_\Omega \to Re\,M_n(\mathbb{C})$ by

$$M_z(\Delta, \alpha) = D_{Re\,\Gamma_\alpha(z)}^{-\frac{1}{2}} (\Delta * \gamma(\alpha)) D_{Re\,\Gamma_\alpha(z)}^{-\frac{1}{2}}$$

(1.7.33)
$$= \frac{1}{2}\left[\Delta_{ij} \frac{\overline{\Gamma_\alpha(z_i)} + \Gamma_\alpha(z_j)}{(Re\,\Gamma_\alpha(z_i) Re\,\Gamma_\alpha(z_j))^{\frac{1}{2}}}\right]_{i,j=1}^n.$$

Fix $\lambda_0 \in \partial\Omega$ and define $s_z^* : Re\,M_n(\mathbb{C}) \to \mathbb{R}$ by

(1.7.34)
$$s_z^*(\Delta) = \sup_{\alpha \in T_\Omega(\lambda_0)} \lambda_{\max}(M_z(\Delta, \alpha))$$

where $\lambda_{\max}(A)$ denotes the maximum eigenvalue of a self-adjoint matrix A. Note that by continuity of $\alpha \mapsto \lambda_{\max}(M_z(\Delta, \alpha))$ and compactness of $T_\Omega(\lambda_0)$, the supremum in (1.7.34) is obtained and is finite. Thus the set $\alpha_z^*(\Delta) \subset T_\Omega(\lambda_0)$ defined by

(1.7.35) $\qquad \alpha_z^*(\Delta) = \{\alpha \in T_\Omega(\lambda_0) : \lambda_{\max}(M_z(\Delta, \alpha)) = s_z^*(\Delta)\}$

is nonempty. In Lemma 1.7.40 below we shall see that $s_z^*(\Delta)$ is, in fact, independent of λ_0.

PROPOSITION 1.7.36. *If $\Delta \in \operatorname{Re} M_n(\mathbb{C})$ satisfies $\Delta \neq 0$, $\Delta_{ii} = 0$ for all $i = 1, \ldots, n$, and $\Delta_{1j} \leq 0$ for all $j = 2, \ldots, n$, then $0 < s_z^*(\Delta) < \infty$, $I - \frac{1}{s_z^*(\Delta)}\Delta \in \partial S_z(\Omega)$, and each $\alpha \in \alpha_z^*(\Delta)$ is an extremal point of $I - \frac{1}{s_z^*(\Delta)}\Delta$.*

We shall postpone the proof of Proposition 1.7.36 until we have established two preliminary results, Lemmas 1.7.39 and 1.7.40.

We shall say that $\Delta \in \operatorname{Re} M_n(\mathbb{C})$ is a *normalized self-adjoint matrix* if $\Delta_{ii} = 1$ for all $i = 1, \ldots, n$ and $\Delta_{1j} \geq 0$ for all $j = 2, \ldots, n$.

LEMMA 1.7.37. *If $A \in \operatorname{Re} M_n(\mathbb{C})$ is a normalized self-adjoint matrix and*

$$(1.7.38) \qquad A * \gamma(\alpha) \geq 0 \qquad \text{for all } \alpha \in T_\Omega(\lambda_0)$$

then $A > 0$.

PROOF. Suppose A is a normalized self-adjoint matrix satisfying (1.7.38). For $t \in [0, 1]$ let

$$G_t = (1-t)I + tA.$$

By (1.7.38) and the fact that $I * \gamma(\alpha) = D_{\operatorname{Re}\Gamma_\alpha(z)} > 0$ for all α,

$$G_t * \gamma(\alpha) \geq 0 \qquad \text{for all } \alpha \in T_\Omega(\lambda_0) \text{ and all } t \in [0,1].$$

By Proposition 1.7.25 if $G_t > 0$ then $G_t \in S_z(\Omega)$, implying, by Lemma 1.7.13, that $G_t \geq c_z > 0$, where c_z is a constant independent of t. Defining $h : [0,1] \to \mathbb{R}$ by

$$h(t) = \lambda_{\min}(G_t)$$

it follows that if $h(t) > 0$ then $h(t) \geq c_z$. But $h(0) = \lambda_{\min}(I) = 1$, and h is continuous. Consequently, by the intermediate value theorem we must have $h(t) \geq c_z$ for all $t \in [0,1]$. In particular, $c_z \leq h(1) = \lambda_{\min}(A)$, thus $A > 0$ and the proof of Lemma 1.7.37 is complete. $\square$

Lemma 1.7.37 and Proposition 1.7.25 immediately imply the following strengthened version of Proposition 1.7.25:

PROPOSITION 1.7.39. *For each $\lambda_0 \in \partial\Omega$, $A \in S_z(\Omega)$ if and only if A is a normalized self-adjoint matrix and (1.7.26) holds.*

LEMMA 1.7.40. *If $\Delta \in \operatorname{Re} M_n(\mathbb{C})$ satisfies the hypotheses of Proposition 1.7.36, then*

$$0 < \sup\{t > 0 : I - t\Delta \in S_z(\Omega)\} = \frac{1}{s_z^*(\Delta)} < \infty.$$

PROOF. Suppose $t > 0$. By Proposition 1.7.39, $I - t\Delta \in S_z(\Omega)$ if and only if

$$(I - t\Delta) * \gamma(\alpha) \geq 0 \qquad \text{for all } \alpha \in T_\Omega(\lambda_0).$$

Since $I * \gamma(\alpha) = D_{\operatorname{Re}\Gamma_\alpha(z)}$, the above assertion is equivalent to

$$\frac{1}{t} D_{\operatorname{Re}\Gamma_\alpha(z)} \geq \Delta * \gamma(\alpha) \qquad \text{for all } \alpha \in T_\Omega(\lambda_0),$$

which holds, by virtue of the positivity of $D_{\operatorname{Re}\Gamma_\alpha(z)}$, if and only if for all $\alpha \in T_\Omega(\lambda_0)$

$$\frac{1}{t} I \geq D_{\operatorname{Re}\Gamma_\alpha(z)}^{-\frac{1}{2}} (\Delta * \gamma(\alpha)) D_{\operatorname{Re}\Gamma_\alpha(z)}^{-\frac{1}{2}}$$
$$= M_z(\Delta, \alpha),$$

which, in turn, is equivalent to the assertion

$$\frac{1}{t} \geq s_z^*(\Delta).$$

If $s_z^*(\Delta) \leq 0$ then the above computation implies that $I - t\Delta \in S_z(\Omega)$ for all $t > 0$, which contradicts the compactness of $S_z(\Omega)$ (Proposition 1.7.7). Therefore, $s_z^*(\Delta) > 0$. That $s_z^*(\Delta) < \infty$ follows from the continuity of $\alpha \mapsto \lambda_{\max}(M_z(\Delta, \alpha))$ and the compactness of $T_\Omega(\lambda_0)$. We have established that $t > 0$ and $I - t\Delta \in S_z(\Omega)$ if and only if $\frac{1}{t} \geq s_z^*(\Delta)$. It follows that $\sup\{t > 0 : I - t\Delta \in S_z(\Omega)\} = 1/s_z^*(\Delta)$. This completes the proof of Lemma 1.7.40.

We remark that by convexity and compactness of $S_z^*(\Omega)$ and the fact that $I \in S_z(\Omega)$, we have

$$I - t\Delta \in S_z(\Omega) \qquad \text{for all } t \in \left[0, \frac{1}{s_z^*(\Delta)}\right].$$

Also note that Lemma 1.7.40 implies $s_z^*(\Delta)$ is independent of λ_0.

We are now ready to prove Proposition 1.7.36. That $0 < s_z^*(\Delta) < \infty$ and $I - \frac{1}{s_z^*(\Delta)}\Delta \in S_z(\Omega)$ follows from Lemma 1.7.40 and the compactness of $S_z(\Omega)$. To prove that $I - \frac{1}{s_z^*(\Delta)}\Delta \in \partial S_z(\Omega)$ and that each $\alpha \in \alpha_z^*(\Omega)$ is an extremal point of $I - \frac{1}{s_z^*(\Delta)}\Delta$, fix $\alpha^* \in \alpha_z^*(\Delta)$, so that

$$s_z^*(\Delta) = \lambda_{\max}(M_z(\Delta, \alpha^*)).$$

There exists a nonzero vector $U \in \mathbb{C}^n$ such that

$$\begin{aligned}
0 &= (s_z^*(\Delta)I - M_z(\Delta, \alpha^*))U \\
&= D_{\operatorname{Re}\Gamma_{\alpha^*}(z)}^{-\frac{1}{2}}(s_z^*(\Delta)D_{\operatorname{Re}\Gamma_{\alpha^*}(z)} - \Delta * \gamma(\alpha^*))D_{\operatorname{Re}\Gamma_{\alpha^*}(z)}^{-\frac{1}{2}}U \\
&= D_{\operatorname{Re}\Gamma_{\alpha^*}(z)}^{-\frac{1}{2}}((s_z^*(\Delta)I - \Delta) * \gamma(\alpha^*))D_{\operatorname{Re}\Gamma_{\alpha^*}(z)}^{-\frac{1}{2}}U.
\end{aligned}$$

But $D_{\operatorname{Re}\Gamma_{\alpha^*}(z)}^{-\frac{1}{2}}$ is nonsingular, hence defining

$$u = D_{\operatorname{Re}\Gamma_{\alpha^*}(z)}^{-\frac{1}{2}}U$$

we have $u \neq 0$ and

$$\left(I - \frac{1}{s_z^*(\Delta)}\Delta\right) * \gamma(\alpha^*)u = 0.$$

Therefore, α^* is an extremal point of $I - \frac{1}{s_z^*(\Delta)}\Delta$. By Lemma 1.7.32 $I - \frac{1}{s_z^*(\Delta)}\Delta \in \partial S_z(\Omega)$. This concludes the proof of Proposition 1.7.36. □

The following result is the converse of Proposition 1.7.36.

PROPOSITION 1.7.41. *For each $G \in \partial S_z(\Omega)$ there exists $\Delta \in \operatorname{Re} M_n(\mathbb{C})$, satisfying $\Delta \neq 0$, $\Delta_{ii} = 0$ for all $i = 1, \ldots, n$, and $\Delta_{1j} \leq 0$ for all $j = 2, \ldots, n$, such that $G = I - \frac{1}{s_z^*(\Delta)}\Delta$.*

PROOF. Fix $G \in \partial S_z(\Omega)$ and let

$$\Delta = I - G.$$

Since G is a normalized positive matrix, $\Delta_{ii} = 0$ for all $i = 1, \ldots, n$ and $\Delta_{1j} \leq 0$ for $j = 2, \ldots, n$. Also, by continuity and positivity of $\alpha \mapsto \operatorname{Re}\Gamma_\alpha(z)$ we have $g_z(I) > 0$,

where g_z is defined by (1.7.21). Hence, $I \notin \partial S_z$, implying $\Delta \neq 0$. Therefore, we are done once we establish that $s_z^*(\Delta) = 1$. Setting
$$t_0 = \sup\{t > 0 : I - t\Delta \in S_z(\Omega)\},$$
we know $t_0 \geq 1$, since $I - \Delta = G \in S_z(\Omega)$, and $I - t_0\Delta \in S_z(\Omega)$ by Proposition 1.7.36 and and Lemma 1.7.40. It is easily verified that the function g_z satisfies
$$g_z(tM) = tg_z(M) \qquad \text{for all } t \geq 0$$
$$\text{and all } M \in \operatorname{Re} M_n(\mathbb{C})$$
and is concave—i.e.,
$$g_z(tM_1 + (1-t)M_2) \geq tg_z(M_1) + (1-t)g(M_2)$$
$$\text{for all } t \in [0,1]$$
$$\text{and all } M_1, M_2 \in \operatorname{Re} M_n(\mathbb{C}).$$

Consequently, if $t_0 > 1$ then
$$g_z(G) = g_z\left(\left(1 - \frac{1}{t_0}\right)I + \frac{1}{t_0}(I - t_0\Delta)\right)$$
$$\geq \left(1 - \frac{1}{t_0}\right)g_z(I) + \frac{1}{t_0}g_z(I - t_0\Delta)$$
$$\geq \left(1 - \frac{1}{t_0}\right)g_z(I) \quad (\text{since } I - t_0\Delta \in S_z(\Omega))$$
$$= \left(1 - \frac{1}{t_0}\right) \inf_{\alpha \in T_\Omega} \lambda_{\min}(D_{\operatorname{Re}\Gamma_\alpha(z)})$$
$$= \left(1 - \frac{1}{t_0}\right) \inf\{\operatorname{Re}\Gamma_\alpha(z_j) : \alpha \in T_\Omega, j = 1,\ldots,n\}$$
$$> 0.$$

But the above strict inequality contradicts $G \in \partial S_z(\Omega)$. Thus it must be the case that $t_0 = 1$. By Lemma 1.7.40 we have $s_z^*(\Delta) = 1$. This completes the proof of Proposition 1.7.41 □

The following result sums up the consequences of Theorem 1.7.24, and Propositions 1.7.36 and 1.7.41.

THEOREM 1.7.42. $S_z^0(\Omega) = \mathcal{D}_z^0(\Omega)$ if and only if for all $\Delta \in \operatorname{Re} M_n(\mathbb{C})$ satisfying $\Delta \neq 0$, $\Delta_{ii} = 0$ for all $i = 1,\ldots,n$ and $\Delta_{1j} \leq 0$ for all $j = 2,\ldots,n$, $I - \frac{1}{s_z^*(\Delta)}\Delta \in D_z(\Omega)$.

1.7.4. Reduction of the rational dilation condition for grammians which are extremal with respect to the spectral set condition. According to Proposition 1.7.7, checking the rational dilation condition for a normalized grammian G is equivalent to determining whether the feasible set of a continuously-parameterized system of linear matrix inequalities is nonempty. There is a well-developed literature devoted to computational techniques for finite systems of linear matrix inequalities [**NN94, VB96**]. Hence, by discretizing the continuous criterion of Proposition 1.7.7 one obtains an approximate, albeit computationally effective, means for checking whether $G \in D_z(\Omega)$.

In Theorem 1.7.24 we established that it suffices to consider grammians $G \in \partial S_z(\Omega)$ when probing the question $S_z^0(\Omega) \stackrel{?}{=} \mathcal{D}_z^0(\Omega)$. The principal result of this section, Theorem 1.7.56, will show that for such extremal grammians the dilation condition can be significantly simplified beyond the form given in Proposition 1.7.7. A special case of this reduction is given in Theorem 1.7.80, which, for $n = 3$ provides a particularly straightforward necessary and sufficient criterion for dilation that obviates linear matrix inequality techniques altogether.

In Section 1.7.3 we introduced the concepts of extremal points and extremal vectors of a grammian $G \in \partial S_z(\Omega)$ and, in Proposition 1.7.36, presented a method for manufacturing such grammians as well as their associated extremal points. Note that once it is known that α^* is an extremal point of G, an extremal vector u^* of G at α^* can be derived by solving the linear vector equation

$$(\gamma(\alpha^*) * G) u^* = 0.$$

The following result gives an important relationship between extremal vectors and the dilation condition given in part (2) of Proposition 1.7.7.

LEMMA 1.7.43. *Suppose $G \in \partial S_z(\Omega) \cap D_z(\Omega)$, let u^* be an extremal vector of G at α^*, and let $\Psi \in (M_n(\mathbb{C}))^m$ be as in part (2) of Proposition 1.7.7.*
The following equations obtain:

$$(1.7.44) \qquad \left(h(\alpha_r^*) * G + \sum_{s=1}^m \varphi_s(\alpha_r^*) \Psi_s \right) u^* = 0, \qquad r = 1, \ldots, m.$$

Moreover, these equations determine $\Psi_1 u^*, \ldots, \Psi_m u^*$ uniquely.

PROOF. By assumption

$$(\gamma(\alpha^*) * G) u^* = 0$$

and

$$h(\lambda) * G + \sum_{r=1}^m \varphi_r(\lambda) \Psi_r \geq 0 \qquad \text{for all } \lambda \in \partial \Omega.$$

The equations (1.7.44) now obtain by invoking formula (1.7.18), which yields

$$\left(\sum_{r=0}^m w_r(\alpha^*) \left(h(\alpha_r^*) * G + \sum_{s=1}^m \varphi_s(\alpha_r^*) \Psi_s \right) \right) u^* = (\gamma(\alpha^*) * G) u^* = 0,$$

and by noting that the weights $w_0(\alpha^*), \ldots, w_m(\alpha^*)$ are positive. (1.7.44) can be rewritten as

$$[\phi_s(\alpha_r^*)]_{r,s=1}^m \begin{pmatrix} \Psi_1 u^* \\ \vdots \\ \Psi_m u^* \end{pmatrix} = - \begin{pmatrix} h(\alpha_1^*) * Gu^* \\ \vdots \\ h(\alpha_m^*) * Gu^* \end{pmatrix}.$$

Existence and uniqueness of vectors $\Psi_1 u^*, \ldots, \Psi_m u^*$ solving the above equation follows from the invertibility of $[\phi_s(\alpha_r^*)]_{r,s=1}^m$, which is implied by the invertibility of the matrix $B(\alpha)$ in Lemma 1.3.20. □

To rephrase the above result, if $G \in \partial S_z(\Omega) \cap D_z(\Omega)$, then $\Psi_1, \ldots, \Psi_m$ are determined uniquely on each extremal vector of G. The power of this observation is doubled by the following result, which implies that extremal vectors never occur in isolation. Recall the definition of $\tau_t(\alpha) \in T_\Omega$ from (1.7.27).

1.7. AN APPLICATION

LEMMA 1.7.45. *Let $G \in \partial S_z$ and let u^* be an extremal vector of G at α^*. For all $t \in \mathbb{R}$, the vector*

$$(1.7.46) \qquad (D_{\Gamma_{\alpha^*}(z)} - it)u^*$$

is an extremal vector of G at $\tau_t(\alpha^*)$. Furthermore, the subspace $[u^*, D_{\Gamma_{\alpha^*}(z)}u^*]$ of $\mathbb{C}^n$ spanned by the locus of extremal vectors of the form (1.7.46) is two-dimensional.

PROOF. That $\gamma(\alpha^*) * Gu^* = 0$ implies $(\gamma(\tau_t(\alpha^*)) * G)(D_{\Gamma_{\alpha^*}(z)} - it)u^* = 0$ follows from formula (1.7.30). Hence, for all $t \in \mathbb{R}$, (1.7.46) is an extremal vector for G at $\tau_t(\alpha^*)$.

The statement that $[u^*, D_{\Gamma_{\alpha^*}(z)}u^*]$ is two-dimensional is equivalent to the assertion that u^* and $D_{\Gamma_{\alpha^*}(z)}u^*$ are not collinear over $\mathbb{C}$, which we shall prove by contradiction. Accordingly, assume that there exists $c \in \mathbb{C}$ such that

$$D_{\Gamma_{\alpha^*}(z)}u^* = cu^*,$$

or, equivalently,

$$\Gamma_{\alpha^*}(z_j)u_j^* = cu_j^* \qquad \text{for all } j = 1, \ldots, n.$$

Because $u^* \neq 0$ and $\operatorname{Re} \Gamma_{\alpha^*} > 0$, we must have $\operatorname{Re} c > 0$. In addition, since $G > 0$, $\langle Gu^*, u^* \rangle > 0$. But then since u^* is an extremal vector of G at α^*,

$$0 = \langle (\gamma(\alpha^*) * G) u^*, u^* \rangle$$

$$= \frac{1}{2} \sum_{k,j=1}^n G_{kj}(\overline{\Gamma_{\alpha^*}(z_k)} + \Gamma_{\alpha^*}(z_j)) \overline{u}_k^* u_j^*$$

$$= \frac{1}{2} \sum_{k,j=1}^n G_{kj}(\bar{c}\overline{u}_k^* u_j^* + c\overline{u}_k^* u_j^*)$$

$$= (\operatorname{Re} c)\langle Gu^*, u^* \rangle > 0.$$

This contradiction completes the proof of Lemma 1.7.45. $\square$

Summing up the consequences of Lemma 1.7.43 and 1.7.45, if $G \in \partial S_z(\Omega)$ and u^* is an extremal vector of G at α^*, a necessary condition for the existence of $\Psi = (\Psi_1, \ldots, \Psi_m)$ satisfying the dilation condition (1.7.8) is that $\Psi_1, \ldots, \Psi_m$ solve

$$(1.7.47) \qquad \left(h(\tau_t(\alpha^*)_r) * G + \sum_{s=1}^m \varphi_s(\tau_t(\alpha^*)_r)\Psi_s\right) u_t^* = 0 \qquad \text{and for all } t \in \mathbb{R}^*,$$

where $u_\infty^* = u^*$ and u_t^* is given by (1.7.46) when $t \neq \infty$. In particular, these equations determine $\Psi_1, \ldots, \Psi_m$ uniquely on the two-dimensional subspace

$$[u^*, D_{\Gamma_{\alpha^*}(z)}u^*] = \operatorname{span}_\mathbb{C}\{u_t^* : t \in \mathbb{R}^*\}.$$

The above observations suggest a practical technique for checking whether a given $G \in \mathcal{S}_z(\Omega)$ belongs to $\Delta_z(\Omega)$. Namely, we select two equations among (1.7.47) and solve for $\Psi_1, \ldots, \Psi_m$ on $[u^*, D_{\Gamma_{\alpha^*}(z)}u^*]$, and then employ semidefinite programming techniques to attempt to solve for self-adjoint $\Psi_1, \ldots, \Psi_m$ on the orthogonal complement of $[u^*, D_{\Gamma_{\alpha^*}(z)}u^*]$ in such a way that (1.7.8) is satisfied. Of course, the infinitude of equations in (1.7.47) overdetermine $\Psi_1, \ldots, \Psi_m$ on $[u^*, D_{\Gamma_{\alpha^*}(z)}u^*]$. Indeed, the consistency among these equations is itself a necessary condition for rational dilation, and will appear as condition (1.7.57) in Theorem 1.7.56 below. These ideas will now be expanded into a formal algorithm, culminating in Theorem 1.7.56.

Throughout the remainder of this section we shall fix $n \in \mathbb{Z}$, $z \in \mathbb{C}^n$, and $G \in M_n(\mathbb{C})$ and assume the following hypotheses:

(1.7.48) $\quad n \geq 3, z = (z_1, \ldots, z_n)$ with $z_1, \ldots, z_n \in \Omega$ distinct,
$G \in \partial S_z(\Omega)$, and u^* is an extremal vector of G at α^*.

We begin by selecting two equations among (1.7.47) and defining a particular solution $\Psi^0 \in (M_n(\mathbb{C}))^m$ to these equations as follows. Let

$$\beta^* = \tau_0(\alpha^*)$$

—that is, β^* is the unique element of T_Ω such that

$$\Gamma_{\alpha^*}(\beta_r^*) = 0 \quad \text{for all } r = 0, \ldots, m.$$

We demand that $\Psi_1^0, \ldots, \Psi_m^0$ satisfy

(1.7.49) $\quad \left(h(\alpha_r^*) * G + \sum_{s=1}^{m} \varphi_s(\alpha_r^*) \Psi_s^0 \right) u^* = 0 \quad r = 1, \ldots, m,$

and

(1.7.50) $\quad \left(h(\beta_r^*) * G + \sum_{s=1}^{m} \varphi_s(\beta_r^*) \Psi_s^0 \right) (D_{\Gamma_{\alpha^*}(z)} u^*) = 0 \quad r = 1, \ldots, m.$

The above equations are special cases of (1.7.47) with $t = \infty$ and $t = 0$ respectively. Recall we are not assuming, *a priori*, that $G \in D_z(\Omega)$. Nevertheless, we may, as in the proof of Lemma 1.7.43, invoke the nonsingularity of $[\phi_s(\alpha_r)]_{r,s=1}^{m}$ for any choice of $\alpha_r \in \partial_r$, $r = 1, \ldots, m$ to conclude that (1.7.49) and (1.7.50) uniquely determine $\Psi_1^0, \ldots, \Psi_m^0$ on the two-dimensional subspace

$$W = [u^*, D_{\Gamma_{\alpha^*}(z)} u^*].$$

We now need to extend $\Psi_1^0, \ldots, \Psi_m^0$ to all of $\mathbb{C}^n$. Accordingly, let $\{b_1, b_2\}$ be an orthonormal basis for W and let $\{b_3, \ldots, b_n\}$ be an orthonormal basis for

$$W^\perp = \mathbb{C}^n \ominus W,$$

the orthogonal complement of W. The vectors b_1 and b_2 can be expressed in terms of u^* and $D_{\Gamma_{\alpha^*}(z)} u^*$ by Gram–Schmidt orthogonalization, thus from (1.7.49) and (1.7.50) one can compute the quantities $\langle \Psi_r^0 b_1, b_j \rangle$ and $\langle \Psi_r^0 b_2, b_j \rangle$, $r = 1, \ldots, m$, $j = 1, \ldots, n$. Extend each Ψ_r^0 to all of $\mathbb{C}^n$ by defining its matrix $[\Psi_r^0]_B$ with respect to the basis $B = (b_1, b_2, \ldots, b_n)$ as follows:

(1.7.51) $\quad [\Psi_r^0]_B = \begin{pmatrix} \langle \Psi_r^0 b_1, b_1 \rangle & \langle \Psi_r^0 b_2, b_1 \rangle & \overline{\langle \Psi_r^0 b_1, b_3 \rangle} & \cdots & \overline{\langle \Psi_r^0 b_1, b_n \rangle} \\ \langle \Psi_r^0 b_1, b_2 \rangle & \langle \Psi_r^0 b_2, b_2 \rangle & \overline{\langle \Psi_r^0 b_2, b_3 \rangle} & \cdots & \overline{\langle \Psi_r^0 b_2, b_n \rangle} \\ \langle \Psi_r^0 b_1, b_3 \rangle & \langle \Psi_r^0 b_2, b_3 \rangle & & & \\ \vdots & & & 0 & \\ \langle \Psi_r^0 b_1, b_n \rangle & \langle \Psi_r^0 b_2, b_n \rangle & & & \end{pmatrix}.$

Equivalently, one can define $\Psi_1^0, \ldots, \Psi_m^0$ in terms of a tuple $\chi^0 \in (\mathcal{L}(\mathbb{C}^n))^m$ determined by

$$\left(h(\alpha_r^*) * G + \sum_{s=1}^{m} \varphi_s(\alpha_r^*) \chi_s^0 \right) u^* = 0 \quad r = 1, \ldots, m,$$

1.7. AN APPLICATION

$$\left(h(\beta_r^*) * G + \sum_{s=1}^{m} \varphi_s(\beta_r^*)\chi_s^0\right)(D_{\Gamma_{\alpha^*}(z)}u^*) = 0 \qquad r = 1,\ldots,m,$$

and

$$\chi_1^0|_{W^\perp} = \chi_2^0|_{W^\perp} = \cdots = \chi_m^0|_{W^\perp} = 0,$$

by setting

$$\Psi_r^0 = \chi_r^0 + (\chi_r^0 - P_W\chi_r^0)^* \qquad r = 1,\ldots,m,$$

where P_W denotes the orthogonal projection onto W. Either of the above constructions results in the same uniquely defined m-tuple of operator $\Psi^0 = (\Psi_1^0,\ldots,\Psi_m^0)$.

We now define $\epsilon_1 : \partial\Omega \to W$, $e_1 : \partial\Omega \to W$, $\epsilon_2 \in W$, $e_2 : \partial\Omega \to W$, $s : \partial\Omega \to \mathbb{R}$, $U : \partial\Omega \to W^\perp$ as follows. For $\lambda \in \partial\Omega$ let

$$\epsilon_1(\lambda) = \begin{cases} u^* & \text{if } \lambda = \alpha_r^*, r = 0,\ldots,m \\ \frac{1}{i}(D_{\Gamma_{\alpha^*}(z)}u^* - \Gamma_{\alpha^*}(\lambda)u^*) & \text{otherwise,} \end{cases}$$

(1.7.52)
$$e_1(\lambda) = \frac{\epsilon_1(\lambda)}{\|\epsilon_1(\lambda)\|},$$

$$\epsilon_2 = D_{\Gamma_{\alpha^*}(z)}u^* + u^*,$$

$$e_2(\lambda) = \frac{\epsilon_2 - \langle\epsilon_2, e_1(\lambda)\rangle e_1(\lambda)}{\|\epsilon_2 - \langle\epsilon_2, e_1(\lambda)\rangle e_1(\lambda)\|}$$

(1.7.53)
$$s(\lambda) = \langle\left(h(\lambda) * G + \sum_{r=1}^{m}\varphi_r(\lambda)\Psi_r^0\right)e_2(\lambda), e_2(\lambda)\rangle,$$

and

(1.7.54)
$$U(\lambda) = P_{W^\perp}\left(h(\lambda) * G + \sum_{r=1}^{m}\varphi_r(\lambda)\Psi_r^0\right)e_2(\lambda),$$

where $P_{W^\perp}$ denotes the orthogonal projection onto $W^\perp$. Note that e_1, e_2, S, and U are real analytic on $\partial\Omega$. Note also, by the fact that $\operatorname{Re} H_\lambda(z)$ is real analytic in λ, and by the Cauchy–Riemann equations, if $w_1, w_2 \in \Omega$ then $\lambda \mapsto \operatorname{Im} H_\lambda(w_1) - \operatorname{Im} H_\lambda(w_2)$ is real analytic on $\partial\Omega$. It follows that $\lambda \mapsto h(\lambda)$ is real analytic, thus so are the functions s and U. In addition, for all $\lambda \in \partial\Omega$, $\{e_1(\lambda), e_2(\lambda)\}$ is an orthonormal basis for W. Observe that as λ ranges through $\partial\Omega$, $i\epsilon_1(\lambda)$ ranges through all extremal vectors in the set $\{u_t^* : t \in \mathbb{R}^*\}$, where $u_\infty^* = u^*$ and u_t^* is given by (1.7.46) when $t \neq \infty$. Therefore, for all λ, $e_1(\lambda)$ is a normalized extremal vector of G. Finally, for each $r \in \{0,\ldots,m\}$ and each $\lambda \in \partial_r$ define

(1.7.55)
$$i_{G,z}(\lambda) = \begin{cases} P_{W^\perp}h(\lambda) * G|_{W^\perp} & \text{if } s \equiv 0 \\ & \text{and } U \equiv 0 \text{ on } \partial_r \\ P_{W^\perp}h(\lambda) * G|_{W^\perp} \\ \quad - \frac{1}{s(\lambda)}U(\lambda) \otimes U(\lambda)|_{W^\perp} & \text{if } s(\lambda) \neq 0 \\ P_{W^\perp}h(\lambda) * G|_{W^\perp} \\ \quad - \lim_{\substack{\mu \to \lambda \\ \mu \in \partial_r}} \frac{1}{s(\mu)} U(\mu) \otimes U(\mu)|_{W^\perp} & \text{if } s \not\equiv \text{ on } \partial_r \\ & \text{and } s(\lambda) = 0 \\ \text{undefined} & \text{if } s \equiv 0 \\ & \text{and } U \not\equiv 0 \text{ on } \partial_r. \end{cases}$$

Because the functions $h(\lambda)$, s, and U are real analytic on $\partial\Omega$, it is straightforward to verify that on each boundary component ∂_r, $r = 0, \ldots, m$, one of the following cases applies.

Case 1: $i_{G,z}$ is defined and real-meromorphic on ∂_r. At points of analyticity, $i_{G,z}$ is $\mathcal{L}(W^\perp)$-valued.

Case 2: $i_{G,z}$ is undefined everywhere on ∂_r. This occurs only when $s \equiv 0$ and $U \not\equiv 0$ on ∂_r.

Note that since $W^\perp$ is a fixed $n-2$ dimensional subspace of $\mathbb{C}^n$, for the purposes of computation we can identify $W^\perp$ with $\mathbb{C}^{n-2}$ and view $i_{G,z}$, where defined, as a map from $\partial\Omega$ into $M_{n-2}(\mathbb{C}^*)$, where $\mathbb{C}^*$ is the extended complex plane.

With the above construction in hand we are now ready to state the main result of this section.

THEOREM 1.7.56. *Under the hypotheses* (1.7.48), $G \in D_z(\Omega)$ *if and only if*

$$(1.7.57) \qquad \left(h(\lambda) * G + \sum_{r=1}^m \varphi_r(\lambda)\Psi_r^0 \right) e_1(\lambda) = 0 \qquad \text{for all } \lambda \in \partial\Omega$$

and

$$(1.7.58) \qquad i_{G,z}(\lambda) \text{ is defined for all } \lambda \in \partial\Omega \text{ and there exist}$$

$$\text{self-adjoint } \Lambda_1, \ldots, \Lambda_m \in \mathcal{L}(W^\perp) \text{ such that}$$

$$i_{G,z}(\lambda) + \sum_{r=1}^m \phi_r(\lambda)\Lambda_r \geq 0 \text{ for all } \lambda \in \partial\Omega.$$

Note the similarity between (1.7.58) and the rational dilation criterion given by part (2) of Proposition 1.7.7. The difference is that (1.7.58) is significantly simpler since the unknowns $\Lambda_1, \ldots \Lambda_m$ are elements of $\mathcal{L}(W^\perp)$ (which is isomorphic to $M_{n-2}(\mathbb{C})$), whereas in Proposition 1.7.7 $\Psi_1, \ldots, \Psi_m$ are elements of $M_n(\mathbb{C})$.

The proof of Theorem 1.7.56 will hinge upon Lemmas 1.7.59–1.7.68 below, in each of which we shall assume that n, z, and G are fixed and satisfy the hypotheses (1.7.48).

LEMMA 1.7.59. *Suppose* $G \in D_z(\Omega)$, *and let* $\Psi \in (M_n(\mathbb{C}))^m$ *be as in part* (2) *of Proposition 1.7.7. There exists* $\chi \in (M_n(\mathbb{C}))^m$ *such that*

(1) $\chi_1, \ldots, \chi_m$ *are self-adjoint*,
(2) $\chi_r|_W = 0$ *for all* $r = 1, \ldots, m$, *and*
(3) $\Psi = \Psi^0 + \chi$.

PROOF. Assuming the existence of $\Psi \in (M_n(\mathbb{C}))^m$ as in Proposition 1.7.7, let

$$\chi_r = 0_W \oplus P_{W^\perp} \Psi_r|_{W^\perp} \qquad r = 1, \ldots, m,$$

where 0_W denotes the zero operator on W and $P_{W^\perp}$ denotes the orthogonal projection onto $W^\perp$. (1) follows form the self-adjointness of $\Psi_1, \ldots, \Psi_m$ and (2) is immediate. To establish (3) note that by Lemmas 1.7.43 and 1.7.45, and the definition of Ψ^0, both Ψ and Ψ^0 satisfy equations (1.7.47) with $t = \infty$ and $t = 0$. Since, by the proof of Lemma 1.7.43, for all r these equations uniquely determine Ψ_r and Ψ_r^0 on the vectors u^* and $D_{\Gamma_{\alpha^*}(z)}u^*$, we have

$$(1.7.60) \qquad \Psi_r|_W = \Psi_r^0|_W = (\Psi_r^0 + \chi_r)|_W, \qquad r = 1, \ldots, m.$$

By (1.7.51) and self-adjointness of Ψ_r we thus have $P_W \Psi_r^0|_{W^\perp} = P_W \Psi_r|_{W^\perp}$. Therefore

(1.7.61) $$\Psi_r|_{W^\perp} = P_W \Psi_r|_{W^\perp} + P_{W^\perp} \Psi_r|_{W^\perp}$$
$$= (\Psi_r^0 + \chi_r)|_{W^\perp}, \qquad r = 1, \ldots, m.$$

since $P_{W^\perp} \Psi_r^0|_{W^\perp} = 0$ and $P_W \chi_r|_{W^\perp} = 0$. Part (3) now follows from (1.7.60) and (1.7.61). □

LEMMA 1.7.62. *If $G \in D_z(\Omega)$ then (1.7.57) obtains.*

PROOF. Assume $G \in D_z(\Omega)$. First note that using Lemma 1.7.43 and 1.7.45 the following assertion is easily established.

(1.7.63)   For every $r \in \{0, \ldots, m\}$ and every $\lambda \in \partial_r$, $e_1(\lambda)$
is an extremal vector of G at $\tau_t(\alpha^*)$, where t is the
unique extended real number such that $(\tau_t(\alpha^*))_r = \lambda$.

Let $\chi = (\chi_1, \ldots, \chi_n)$ be as in Lemma 1.7.59 and let $\Psi = \Psi^0 + \chi$. Fix $r_0 \in \{0, \ldots, m\}$ and $\lambda \in \partial_{r_0}$, and let $t \in \mathbb{R}^*$ satisfy $(\tau_t(\alpha^*))_{r_0} = \lambda$. By assertion (1.7.63) and Lemma 1.7.43,

$$\left(h((\tau_t(\alpha^*))_r) * G + \sum_{s=1}^m \varphi_s((\tau_t(\alpha^*))_r) \Psi_s \right) e_1(\lambda) = 0 \qquad \text{for all } r = 1, \ldots, m.$$

But by formula (1.7.18) and the fact that $\gamma(\tau_t(\alpha^*)) * G e_1(\lambda) = 0$, the above equality holds for all $r = 0, 1, \ldots, m$. Setting $r = r_0$ and noting that $\chi_s e_1(\lambda) = 0$ for all $s = 1, \ldots, m$, we have

$$0 = \left(h(\lambda) * G + \sum_{r=1}^m \varphi_r(\lambda) \Psi_r \right) e_1(\lambda)$$
$$= \left(h(\lambda) * G + \sum_{r=1}^m \varphi_r(\lambda) \Psi_r^0 \right) e_1(\lambda),$$

and the proof of Lemma 1.7.62 is complete. □

LEMMA 1.7.64. *If (1.7.57) holds then for all $\lambda \in \partial \Omega$*

(1.7.65) $$P_W \left(h(\lambda) * G + \sum_{r=1}^m \varphi_r(\lambda) \Psi_r^0 \right)|_W \geq 0.$$

PROOF. Assume (1.7.57) holds. We shall first establish the following inequality

(1.7.66) $$\left\langle \left(h(\lambda) * G + \sum_{r=1}^m \varphi_r(\lambda) \Psi_r^0 \right) e_1(\mu), e_1(\mu) \right\rangle \geq 0 \qquad \text{for all } \lambda, \mu \in \partial \Omega.$$

To prove this inequality fix $q, r \in \{0, \ldots, m\}$, $\lambda \in \partial_q$ and $\mu \in \partial_r$. Let $t \in \mathbb{R}^*$ be the unique extended real number such that

$$(\tau_t(\alpha^*))_r = \mu.$$

From (1.7.27) and the definition of e_1 we conclude that

$$e_1((\tau_t(\alpha^*))_s) = e_1(\mu), \qquad s = 0, \ldots, m,$$

and so, by (1.7.57),
(1.7.67)
$$\left(h((\tau_t(\alpha^*))_s) * G + \sum_{r=1}^{m} \varphi_r((\tau_t(\alpha^*))_s)\Psi_r^0\right) e_1(\mu) = 0 \quad \text{for all } s = 0, \ldots, m.$$

Define $\beta \in T_\Omega$ by
$$\beta_s = \begin{cases} (\tau_t(\alpha^*))_s & s \neq q \\ \lambda & s = q. \end{cases}$$

By formula (1.7.18) and (1.7.67),
$$\gamma(\beta) * Ge_1(\mu) = \sum_{s=0}^{m} w_s(\beta) \left(h(\beta_s) * G + \sum_{r=1}^{m} \varphi_r(\beta_s)\Psi_r^0\right) e_1(\mu)$$
$$= w_q(\beta) \left(h(\lambda) * G + \sum_{r=1}^{m} \varphi_r(\lambda)\Psi_r^0\right) e_1(\mu).$$

The inequality (1.7.66) now follows upon recalling our assumption $G \in S_z(\Omega)$, which implies $\gamma(\beta) * G \geq 0$, and the fact that $w_q(\beta) > 0$.

To prove (1.7.65) fix $\lambda \in \partial\Omega$ and let
$$M = P_W \left(h(\lambda) * G + \sum_{r=1}^{m} \varphi_r(\lambda)\Psi_r^0\right)|_W$$

and
$$y_1 = e_1(\lambda),$$
so that $My_1 = 0$ (recall we are assuming (1.7.57)). Define
$$y_2 = \begin{cases} \frac{1}{i}D_{\Gamma_{\alpha^*}(z)}u^* & \text{if } y_1 = u^* \\ u^* & \text{otherwise.} \end{cases}$$

Note that $\frac{1}{i}\Gamma_{\alpha^*}(\mu)$ ranges through $\mathbb{R}^*$ as μ ranges through $\partial\Omega$. Thus, for every $t \in \mathbb{R}^*$ there exists $\mu \in \partial\Omega$ such that $y_1 + ty_2$ is collinear with $e_1(\mu)$ over $\mathbb{C}$. Therefore, by (1.7.66), for all $t \in \mathbb{R}$
$$0 \leq \langle M(y_1 + ty_2), y_1 + ty_2 \rangle$$
$$= t\langle My_2, y_1 \rangle + t^2 \langle My_2, y_2 \rangle,$$

implying $\langle My_2, y_1 \rangle = 0$ and $\langle My_2, y_2 \rangle \geq 0$. Since $\text{span}_\mathbb{C}\{y_1, y_2\} = W$ it follows that $M \geq 0$, and the proof of Lemma 1.7.64 is complete. □

LEMMA 1.7.68. *If (1.7.57) holds then* $\Psi_1^0, \ldots, \Psi_m^0$ *are self-adjoint.*

PROOF. Observe that Ψ_r^0 is self-adjoint if and only if $P_W \Psi_r^0|_W$, which is represented by the upper left 2×2 block of (1.7.51), is self-adjoint. Assuming (1.7.57), Lemma 1.7.64 implies that $P_W \left(h(\lambda) * G + \sum_{r=1}^{m} \varphi_r(\lambda)\Psi_r^0\right)|_W$ is nonnegative, hence self-adjoint for all $\lambda \in \partial\Omega$. Formula (1.7.19) yields, for $r = 1, \ldots, m$,
$$P_W \Psi_r^0|_W = \int_{\partial\Omega} P_W \left(h(\lambda) * G + \sum_{s=1}^{m} \varphi_s(\lambda)\Psi_s^0\right)|_W \phi_r(\lambda)\, d\sigma(\lambda),$$

hence $P_W \Psi_r^0|_W$ is self-adjoint.

Proof of Theorem 1.7.56. First we shall establish that if $G \in D_z(\Omega)$ then (1.7.57) and (1.7.58) obtain. Assuming $G \in D_z(\Omega)$, Lemma 1.7.62 implies (1.7.57). To prove (1.7.58) let $\chi \in (M_n(\mathbb{C}))^m$ be as in Lemma 1.7.59, so that

$$(1.7.69) \qquad h(\lambda) * G + \sum_{r=1}^{m} \varphi_r(\lambda) \left(\Psi_r^0 + \chi_r \right) \geq 0 \qquad \text{for all } \lambda \in \partial\Omega.$$

Observe that since $e_1(\lambda), e_2(\lambda) \in W$ for all $\lambda \in \partial\Omega$ and $\chi_r|_W = 0$ for all $r = 1, \ldots, m$, we have

$$(1.7.70) \qquad s(\lambda) = \langle \left(h(\lambda) * G + \sum_{r=1}^{m} \varphi_r(\lambda) \Psi_r^0 \right) e_2(\lambda), e_2(\lambda) \rangle$$
$$= \langle h(\lambda) * G + \sum_{r=1}^{m} \varphi_r(\lambda) \left(\Psi_r^0 + \chi_r \right) e_2(\lambda), e_2(\lambda) \rangle$$

and

$$(1.7.71) \qquad U(\lambda) = P_{W^\perp} \left(h(\lambda) * G + \sum_{r=1}^{m} \varphi_r(\lambda) \Psi_r^0 \right) e_2(\lambda)$$
$$= P_{W^\perp} \left(h(\lambda) * G + \sum_{r=1}^{m} \varphi_r(\lambda) \left(\Psi_r^0 + \chi_r \right) \right) e_2(\lambda).$$

Also, from (1.7.57),

$$(1.7.72) \qquad h(\lambda) * G + \sum_{r=1}^{m} \varphi_r(\lambda) \left(\Psi_r^0 + \chi_r \right) e_1(\lambda) = 0 \qquad \text{for all } \lambda \in \partial\Omega.$$

Next, note that from (1.7.51)

$$(1.7.73) \qquad P_{W^\perp} \Psi_r^0 |_{W^\perp} = 0 \qquad r = 1, \ldots, m.$$

Defining

$$\Lambda_r = P_{W^\perp} \chi_r |_{W^\perp} \qquad r = 1, \ldots, m,$$

we obtain, using (1.7.70), (1.7.71), (1.7.72), and (1.7.73), and invoking the self-adjointness of $h(\lambda) * G + \sum_{r=1}^{m} \varphi_r(\lambda) \left(\Psi_r^0 + \chi_r \right)$, the following block decomposition of $h(\lambda) * G + \sum_{r=1}^{m} \varphi_r(\lambda) \left(\Psi_r^0 + \chi_r \right)$ in terms of $e_1(\lambda), e_2(\lambda)$ and $W^\perp$:

$$(1.7.74) \qquad [h(\lambda) * G + \sum_{r=1}^{m} \varphi_r(\lambda) \left(\Psi_r^0 + \chi_r \right)]_{(e_1(\lambda), e_2(\lambda), W^\perp)} = \begin{pmatrix} 0 & 0 & 0 \ldots 0 \\ \hline 0 & s(\lambda) & U(\lambda)^* \\ \hline 0 & & P_{W^\perp} h(\lambda) * G|_{W^\perp} \\ \vdots & U(\lambda) & \\ 0 & & + \sum_{r=1}^{m} \phi_r(\lambda) \Lambda_r \end{pmatrix}.$$

Recall that we must prove that for all $\lambda \in \partial\Omega$, $i_{G,z}(\lambda)$ is defined and

$$(1.7.75) \qquad i_{G,z}(\lambda) + \sum_{r=1}^{m} \phi_r(\lambda) \Lambda_r \geq 0.$$

We shall show, in fact, that $i_{G,z}$ is real-analytic on $\partial\Omega$ and (1.7.75) holds on each boundary component ∂_r, $r = 0, \ldots, m$.

From (1.7.69) and (1.7.74) we see that the function s is nonnegative on $\partial\Omega$, and that if $s(\lambda) = 0$, then $U(\lambda) = 0$. Because s is real-analytic on $\partial\Omega$, on each boundary component ∂_r, $r = 0, \ldots, m$ there are two possibilities.

Case 1: $s \equiv 0$ on ∂_r. This forces $U \equiv 0$ on ∂_r. From (1.7.55) we have $i_{G,z}(\lambda) = P_{W^\perp} h(\lambda) * G|_{W^\perp}$. In particular, $i_{G,z}$ is real-analytic on ∂_r and from (1.7.69) and (1.7.74) we obtain (1.7.75) on ∂_r.

Case 2: s has at most finitely many zeros on ∂_r and s has positive values at all other points of ∂_r. Let $\mathcal{Z} \subseteq \partial_r$ be the set of zeros of s on ∂_r. (1.7.55) yields, for all $\lambda \in \partial_r/\mathcal{Z}$

$$(1.7.76) \qquad i_{G,z}(\lambda) = P_{W^\perp} h(\lambda) * G|_{W^\perp} - \frac{1}{s(\lambda)} U(\lambda) \otimes U(\lambda).$$

In particular, $i_{G,z}$ is real-meromorphic on ∂_r and from (1.7.69) and (1.7.74) we obtain (1.7.75) on $\partial_r/\mathcal{Z}$. But $i_{G,z}$ is, in fact, bounded on $\partial_r/\mathcal{Z}$; this can be seen from (1.7.75) and (1.7.76), which imply, for all $\lambda \in \partial_r/\mathcal{Z}$, the inequalities

$$i_{G,z}(\lambda) \geq -\sum_{r=1}^{m} \phi_r(\lambda) \Lambda_r$$

and

$$i_{G,z}(\lambda) \leq P_{W^\perp} h(\lambda) * G|_{W^\perp}.$$

Therefore, $i_{G,z}$ is real-analytic on ∂_r, and thus the inequality (1.7.75) extends to all of ∂_r by continuity.

In summary, we have shown that on each ∂_r, $r = 0, \ldots, m$, $i_{G,z}$ is real analytic and (1.7.75) holds. This completes the proof that $G \in D_z(\Omega)$ implies (1.7.57) and (1.7.58).

We shall now assume (1.7.57) and (1.7.58) and shall prove that $G \in D_z(\Omega)$. By Proposition 1.7.7 and Lemma 1.7.68, it suffices to show that there exists $\chi \in (M_n(\mathbb{C}))^m$ such that $\chi_1, \ldots, \chi_m$ are self-adjoint and for all $\lambda \in \partial\Omega$

$$(1.7.77) \qquad h(\lambda) * G + \sum_{r=1}^{m} \varphi_r(\lambda) \left(\Psi_r^0 + \chi_r \right) \geq 0.$$

For each $r = 1, \ldots, m$ define

$$\chi_r = 0_W \oplus \Lambda_r$$

where 0_W denotes the zero operator on the subspace W. Since $\Lambda_1, \ldots, \Lambda_m$ are self-adjoint, and, by Lemma 1.7.68, $\Psi_1^0, \ldots, \Psi_m^0$ are self-adjoint,

$$h(\lambda) * G + \sum_{r=1}^{m} \varphi_r(\lambda) \left(\Psi_r^0 + \chi_r \right)$$

is self-adjoint for all $\lambda \in \partial\Omega$. Consequently, $h(\lambda) * G + \sum_{r=1}^{m} \varphi_r(\lambda) \left(\Psi_r^0 + \chi_r \right)$ admits the block decomposition (1.7.74). By Lemma 1.7.64, $s(\lambda) \geq 0$ for all $\lambda \in \partial\Omega$. By real-analyticity of s, on each boundary component ∂_r, $r = 1, \ldots, m$ there are two possibilities.

Case 1: $s \equiv 0$ on ∂_r. According to (1.7.55), our assumption that $i_{G,z}(\lambda)$ is defined on ∂_r implies $U \equiv 0$ on ∂_r, so that $i_{G,z}(\lambda) = P_{W^\perp} h(\lambda) * G|_{W^\perp}$ for all $\lambda \in \partial_r$. Therefore, from (1.7.58) and (1.7.74) we obtain (1.7.77) for all $\lambda \in \partial_r$.

Case 2: s has at most finitely many zeros on ∂_r and s has positive values at all other points of ∂_r. Let $\mathcal{Z} \subseteq \partial_r$ denote the set of zeros of s on ∂_r. By (1.7.55) we have, for all $\lambda \in \partial_r/\mathcal{Z}$, $i_{G,z}(\lambda) = P_{W^\perp} h(\lambda) * G|_{W^\perp} - \frac{1}{s(\lambda)} U(\lambda) \otimes U(\lambda)$. This implies by virtue of (1.7.58) and (1.7.74), that the inequality (1.7.77) obtains for all $\lambda \in \partial_r/\mathcal{Z}$. (1.7.77) now extends by continuity to all of ∂_r.

In summary, we have established that (1.7.57) and (1.7.58) imply (1.7.77) on each $\partial_1, \ldots, \partial_m$ and therefore $G \in D_z(\Omega)$. This completes the proof of Theorem 1.7.56. □

We remark that in the proof of Theorem 1.7.56 we established that when (1.7.58) obtains, $i_{G,z}$ is, in fact, real-analytic on $\partial \Omega$.

The following result provides convenient necessary conditions for rational dilation that do not rely on semidefinite programming techniques.

COROLLARY 1.7.78. (Necessary conditions for rational dilation.) *Under the hypotheses* (1.7.48), $G \in D_z(\Omega)$ *only if* (1.7.57) *obtains and*

(1.7.79)
$$i_{G,z}(\lambda) \text{ is defined for all } \lambda \in \partial\Omega \text{ and}$$

$$\sum_{r=0}^{m} w_r(\alpha) i_{G,z}(\alpha_r) \geq 0 \text{ for all } \alpha \in T_\Omega.$$

PROOF. Recall, from (1.3.21), that for all $s = 1, \ldots, m$, $\sum_{r=0}^{m} w_r(\alpha) \phi_s(\alpha_r) = 0$ and, from Theorem 1.3.17, that the weights $w_0(\alpha), \ldots, w_m(\alpha)$ are positive. Therefore, (1.7.58) implies (1.7.79). □

The following result shows that the necessary conditions of the above result become, in fact, necessary and sufficient when $n = 3$ or, equivalently, when $W^\perp$ is one-dimensional and thus $i_{G,z}$ is scalar-valued.

THEOREM 1.7.80. *Under the hypotheses* (1.7.48), *if* $n = 3$, *then* $G \in D_z(\Omega)$ *if and only if* (1.7.57) *and* (1.7.79) *obtain.*

PROOF. By Corollary 1.7.78 we need only establish that (1.7.57) and (1.7.79) imply $G \in D_z(\Omega)$. But this will follow from Proposition 1.5.3 and Theorem 1.7.56 once we establish that (1.7.79) implies $i_{G,z}$ is continuous on $\partial\Omega$. Accordingly, assume (1.7.79). It will be shown that, in fact, $i_{G,z}$ is real-analytic on $\partial\Omega$. By (1.7.79), $i_{G,z}$ is defined on $\partial\Omega$, implying, by (1.7.55) that $i_{G,z}$ is real-meromorphic on $\partial\Omega$. Let $B = \max\{w_r(\alpha) : r \in \{0, \ldots, m\}, \alpha \in T_\Omega\}$ and let $b = \min\{w_r(\alpha) : r \in \{0, \ldots, m\}, \alpha \in T_\Omega\}$. Choose $\alpha \in T_\Omega$ so that $i_{G,z}$ is real-analytic, hence finite, at each $\alpha_0, \ldots, \alpha_m$. Fixing $r_0 \in \{0, \ldots, m\}$ and $\lambda \in \partial_{r_0}$, and replacing α_{r_0} by λ in (1.7.79) we obtain

$$i_{G,z}(\lambda) \geq -\frac{B}{b} \sum_{\substack{r=0 \\ r \neq r_0}}^{M} |i_{G,z}(\alpha_r)|$$

$$\geq -\frac{B}{b} \sum_{r=0}^{m} |i_{G,z}(\alpha_r)|$$

The right hand side of the above inequality is independent of s and λ, implying that $i_{G,z}$ is uniformly bounded below on $\partial\Omega$. But as shown in the proof of Theorem 1.7.56, $i_{G,z}$, being defined on all of $\partial\Omega$, is uniformly bounded above on $\partial\Omega$. Therefore, $i_{G,z}$ is a bounded meromorphic function, and is hence real-analytic on $\partial\Omega$. This completes the proof of Theorem 1.7.80. $\square$

1.7.5. An preliminary algorithm for probing Problem 1.1.4. We shall conclude Section 1.7 by briefly laying out the ideas behind an algorithm for searching for a counterexample in $\mathcal{S}_z^0(\Omega)$ to Problem 1.1.4. By Theorem 1.7.42, if such a counterexample exists then there exists a multiply connected domain $\Omega \subseteq \mathbb{C}$ with analytic boundary, $n \geq 3$, $z = (z_1, \ldots, z_n)$ with $z_1, \ldots, z_n \in \Omega$ distinct, and an element Δ of the set

$$R = \{\Delta \in Re\, M_n(\mathbb{C}) : \Delta_{ii} = 0 \text{ for } i = 1, \ldots, n$$
$$\text{and } \Delta_{1j} \leq 0 \text{ for } j = 2, \ldots, n\}$$

such that $I - \frac{1}{s_z^*(\Delta)}\Delta$ does not belong to $D_z(\Omega)$. Therefore, it would seem that a plausible method for attempting to evolve a counterexample of this form would be to select Ω, $n \geq 3$, and z, and iterate the following steps.

(1) Generate a random candidate $\Delta \in R$.
(2) Fix $\lambda_0 \in \partial\Omega$ and solve for $s_z^*(\Delta)$ in the optimization problem (1.7.34).
(3) By Proposition 1.7.36, any point $\alpha^* \in T_\Omega$ solving (1.7.34) is an extremal point of $I - \frac{1}{s_z^*(\Delta)}\Delta$. Solve the linear equation $(I - \frac{1}{s_z^*(\Delta)}\Delta) * \gamma(\alpha^*)u^* = 0$ for an extremal vector u^*. Solve (1.7.49), (1.7.50), and (1.7.51) for $\Psi^0 \in (M_n(\mathbb{C}))^m$. Formulae for e_1, e_2, s, U and $i_{G,z}$ can now be computed from (1.7.52), (1.7.53), (1.7.54), and (1.7.55).
(4) Check whether $I - \frac{1}{s_z^*(\Delta)}\Delta \in D_z(\Omega)$ using the criteria of Theorem 1.7.80 if $n = 3$, or the criteria of Theorem 1.7.56 together with semidefinite programming techniques if $n \geq 4$. (If a counterexample is found for $n \geq 4$, it may in fact violate criterion (1.7.79), thus circumventing the semidefinite programming required by the more stringent criterion (1.7.58)).

Of course, to implement this algorithm, one must first compute approximates for the functions $\phi_1, \ldots, \phi_m$ and H_λ. An approximate of Γ_α can then be computed by means of formulae (1.3.21) and (1.4.9). This problem is taken up in Chapter 3.

As discussed in the introduction of Section 1.7, computer experiments based on (1)-(4) have produced for various domains Ω, various values of n, and various choices of spectral points $z_1, \ldots, z_n \in \Omega$, numerous examples of extremal grammians $G \in \partial\mathcal{S}_z(\Omega)$, all of which satisfy $G \in \mathcal{D}_z(\Omega)$. The failure to achieve a counterexample to Problem 1.1.4 by this means led the authors to develop an extension of the above algorithm - described in Chapter 2 - which manufactures "hyperextremal" grammians : grammians which meet the spectral set conditions more sharply than do the typical elements of $\partial\mathcal{S}_z(\Omega)$. For $n = 3$, the hyperextremals that we have produced satisfy $G \in \mathcal{D}_z(\Omega)$ (cf. [**Rap02**]). However, for $n = 4$ we have generated a considerable body of empirical evidence that many (perhaps all) such hyperextremals do not belong to $\mathcal{D}_z(\Omega)$, and therefore provide counterexamples to Problem 1.1.4. Thus, the computations which we detail in Chapter 2 strongly suggest a negative answer to Problem 1.1.4 in general (and, in particular when $n \geq 4$). Additionally, the computations suggest a positive answer in the case $n \leq 3$. It is interesting to note that the criterion (1.7.57) of Theorem 1.7.56 is satisfied, to the numerical limits of all calculations, for all extremal and hyperextremal G that we have studied, independent of whether or not $G \in \mathcal{D}_z(\Omega)$ (i.e. whether or not (1.7.58) holds). This observation leads to the additional speculation that (1.7.57) holds for all $G \in \partial \mathcal{S}_z(\Omega)$.

CHAPTER 2

The Computational Generation of Counterexamples to the Rational Dilation Conjecture

2.1. Introduction

Let $\mathcal{H}$ be a complex Hilbert space, and let $\mathcal{L}(\mathcal{H})$ to denote the bounded linear operators from $\mathcal{H}$ into $\mathcal{H}$. Let $\mathbb{D}$ denote the open unit disk in $\mathbb{C}$, the complex plane. A classic result in the theory of linear operators on Hilbert space introduced by von Neumann in [**vN51**] is the following.

VON NEUMANN'S INEQUALITY. *If $T \in \mathcal{L}(\mathcal{H})$, $\|T\| \leq 1$, and p is a polynomial then*
$$\|p(T)\| \leq \max_{z \in \mathbb{D}^-}|p(z)|.$$

To generalize this result, in [**vN51**] von Neumann also introduced the theory of spectral sets. For a compact subset $K \subseteq \mathbb{C}$, let $\operatorname{Rat}(K)$ be the set of rational functions whose poles lie off K.

DEFINITION 2.1.1. *If $T \in \mathcal{L}(\mathcal{H})$, and K is a compact subset of $\mathbb{C}$ such that $\sigma(T) \subseteq K$, then K is a **spectral set** for T provided*

(2.1.2) $$\|f(T)\| \leq \max_{z \in K}|f(z)| \text{ for all } f \in \operatorname{Rat}(K).$$

By Runge's Theorem, von Neumann's inequality is equivalent to the statement that if $\|T\| \leq 1$ then $\mathbb{D}^-$ is a spectral set for T. Conversely, by setting $f(z) = z$ in the definition, we see that if $\mathbb{D}^-$ is a spectral set for T, then $\|T\| \leq 1$. Thus, von Neumann's Inequality can be restated as the assertion that $\mathbb{D}^-$ is a spectral set for T if and only if $\|T\| \leq 1$. Since von Neumann's introduction of the definition of spectral sets, the theory has drawn considerable attention (see [**Foi59**], [**And63**], [**Leb63**], [**Sar65b**], [**Ber68**], [**Mla72**], [**Lau73**], [**Par70**], [**Var74**], [**Arv69**], [**Arv72**], [**DP86**], [**Agl80**], [**Agl85**], [**Agl90**], and [**Pau88**] for a survey). Nevertheless, some basic questions remain unanswered. Most notably, for a given compact set K, one would like to have a simple model for the linear operators that have K as a spectral set.

For some sets K, such models have been obtained via rational dilations. The seminal result of this type is the following theorem of Sz.-Nagy [**SN53**].

Sz.-NAGY DILATION THEOREM. *If $T \in \mathcal{L}(\mathcal{H})$ with $\|T\| \leq 1$, then there exists a Hilbert space $\mathcal{K}$ with $\mathcal{K} \supseteq \mathcal{H}$ and a unitary $U \in \mathcal{L}(\mathcal{K})$ such that if $P_{\mathcal{H}}$ denotes the orthogonal projection of $\mathcal{K}$ onto $\mathcal{H}$, then*

(2.1.3) $$p(T) = P_{\mathcal{H}} p(U)|_{\mathcal{H}},$$

for all polynomials p.

Again, Runge's Theorem implies that the polynomials in (2.1.3) could be replaced by the rational functions with poles off $\mathbb{D}^-$, leading to the following definition.

DEFINITION 2.1.4. *If* $T \in \mathcal{L}(\mathcal{H})$ *with* $\sigma(T) \subseteq K$ *then* T *has a* **rational dilation to** ∂K *provided there exists a Hilbert space* $\mathcal{K} \supseteq \mathcal{H}$ *and a normal operator* $N \in \mathcal{L}(\mathcal{K})$ *with* $\sigma(N) \subseteq \partial K$ *such that if* $P_\mathcal{H}$ *denotes the orthogonal projection of* $\mathcal{K}$ *onto* $\mathcal{H}$, *then*
$$f(T) = P_\mathcal{H} f(N)|_\mathcal{H},$$
for all $f \in Rat(K)$.

The Spectral Theorem implies that if $T \in \mathcal{L}(\mathcal{H})$ has a rational dilation to ∂K, then K is a spectral set for T. Furthermore, the common condition, $\|T\| \leq 1$, in von Neumann's Inequality and the Sz.-Nagy Dilation Theorem, implies that the converse is true in the case $K = \mathbb{D}^-$. Thus, the notions of spectral sets and rational dilation are equivalent when $K = \mathbb{D}^-$. Are the notions of spectral set and rational dilation equivalent for other compact sets $K \subseteq \mathbb{C}$?

Specifically, let Ω be a bounded domain in $\mathbb{C}$ whose boundary $\partial\Omega$ consists of $m+1$ disjoint real analytic Jordan curves, and consider the following conjecture.

RATIONAL DILATION CONJECTURE. *If* $T \in \mathcal{L}(\mathcal{H})$ *and* Ω^- *is a spectral set for* T, *then* T *has a rational dilation to* $\partial\Omega$.

As noted above, the conjecture is true when $\Omega = \mathbb{D}$. Foias, Berger, and Lebow independently proved that the conjecture is true when $R(\Omega^-)$ is a Dirichlet algebra on $\partial\Omega$ ([**Foi59**], [**Leb63**], [**Ber68**]). Sarason proved the conjecture is true when Ω is a domain whose complement is connected [**Sar65b**]. Later, Agler proved that the conjecture is true in the case when Ω is an annulus [**Agl85**]. The conjecture is also true in the case when $dim(\mathcal{H}) = 2$ and Ω is any domain in $\mathbb{C}$ ([**Mis84**], [**Pau87**], [**Agl90**]).

The "simplest" unresolved case, therefore is when Ω is a two-holed domain and $dim(\mathcal{H}) \geq 3$. A number of theoretical tools have been brought to bear on this problem. In [**Arv69**] and [**Arv72**], Arveson restated the problem in terms of contractive and completely contractive representations of $R(K)$. Also, in [**AD76**] Abrahamse and Douglass introduced the *bundle shift* operators, which provide models of operators that have rational dilations to $\partial\Omega$ for a multiply-connected domain Ω. More recently, Ball and Vinnikov [**Vin98**] suggested a construction for operators having Ω^- as a spectral set by using a pair of commuting contractions with finite defects on the Schottky double of Ω. Yet, a definitive resolution of the rational dilation conjecture remains elusive[1].

Here, we present an approach to the rational dilation conjecture based on machine computation. Implementation of this approach yields specific 4×4 matrices that we believe represent counterexamples to the rational dilation conjecture. This same computational approach has failed to produce 3×3 counterexamples, and in fact the approach yields numerical evidence that the 3×3 case may well be true. We limit the discussion in this memoir to the 4×4 case. See [**Rap02**] for further discussion of the 3×3 case.

---

[1]While the present monograph was under review, Dritschel and McCullough [**DM05**] provided a nonconstructive proof of a counterexample to the rational dilation conjecture.

2.1. INTRODUCTION

In Section 2.2, after presenting some requisite function theory, we explicitly describe a computationally efficient criterion for a given matrix $M \in \mathbb{C}^{n,n}$, the $n \times n$ matrices with complex entries, to have a given m-holed domain Ω^- as a spectral set. This criterion has the form of the *feasibility* of a system of linear matrix inequalities continuously parameterized by an m-dimensional torus. Similarly, in Section 2.2, we describe a criterion for M to have a rational dilation to $\partial\Omega$. This criterion has the form of the *solvability* of a system of linear matrix inequalities continuously parameterized by $\partial\Omega$. The linear matrix inequalities introduced give rise to a number of natural theoretical notions that are developed in the remainder of the first chapter of the memoir. Recent developments in the theory of semidefinite programming ([**VB96**], [**NN94**], [**Wri97**]) will allow us to compute with these linear matrix inequalities efficiently. Section 2.2 closes with the introduction of "grammian coordinates" for the matrix M. These coordinates display the matrix M in terms of the grammian of its eigenvectors, and are of great theoretical and computational importance.

In turns out that not all matrices M that have a given set $K = \Omega^-$ as a spectral set are qualitatively the same. Some matrices M are "extremal" in that sense that they meet the inequality (2.1.2) more sharply than others. In Section 2.3 we define such "extremal" matrices, and also define the stronger notion of "hyperextremal" matrices. We also show how to associate a certain linear system $Ax = b$ to a hyperextremal matrix M in the case when $m = 2$ and $n = 4$. The solution of this linear system gives specific estimates of the distance (in grammian coordinates) between the hyperextremal matrix M and the set of matrices that have rational dilations to $\partial\Omega$.

Now, just as not all matrices M that have a given set Ω^- as a spectral set are qualitatively the same, not all matrices that have a rational dilation to $\partial\Omega$ are qualitatively the same. Some matrices M are "dilation extremal" in the sense that the requirements of Definition 2.1.4 are met more sharply than others. In Section 2.4, we give a precise definition of dilation extremal matrices and then show how to associate a certain system of linear inequalities $L_{\lambda,x}(\mathbf{Z}) \geq \eta$ to a dilation extremal matrix M. From the feasibility of this system of linear inequalities, we derive an estimate of the distance (in grammian coordinates) between M and the boundary of the set of matrices that have a rational dilation to $\partial\Omega$. Either of the estimates from Section 2.3 or from Section 2.4 is sufficient to justify a counterexample.

Section 2.5 begins the description of a computer implementation of our approach to the rational dilation conjecture. We present algorithms for the construction of extremal, hyperextremal, and dilation extremal grammians, and also an algorithm for the computation of the necessary functions on Ω^-. In Section 2.6 we present a specific 4×4 matrix obtained via the algorithms of Section 2.5. We then compute the distance estimates derived in Sections 2.3 and 2.4 to give two independent checks that there is a matrix M that has Ω^- as a spectral set, but does not have a rational dilation to $\partial\Omega$. Hence, we obtain a counterexample to the rational dilation conjecture. Finally in Section 2.7, we present further plausibility arguments for the counterexample, and highlight some interesting questions and new discoveries resulting from the computation.

2.2. Mathematical Preliminaries

This section presents a number of preliminary results that form the foundation of our search for a counterexample. We organize these results into three subsections. The first subsection defines certain functions on the domain Ω, the so called *Herglotz kernels of the first and second kinds*. We then present two theorems: the first uses the Herglotz kernel of the first kind to give a condition for a given $M \in \mathbb{C}^{n,n}$ to have Ω^- as a spectral set (Theorem 2.2.5) and the second uses the Herglotz kernel of the second kind to give a condition for M to have a rational dilation to $\partial\Omega$ (Theorem 2.2.7). Unlike Definitions 2.1.1 and 2.1.4, the conditions in these theorems are conducive to machine computation. Furthermore, they lead to a number of useful theoretical ideas that will be explored in the rest of the memoir.

The second subsection introduces two functional identities satisfied by the Herglotz kernels that not only lead to a simplification of the spectral set condition (Proposition 2.2.32), but also give rise to a linear equation associated with the rational dilation condition under certain circumstances (Proposition 2.2.28).

Finally, in the third subsection, grammian coordinates are introduced. These coordinates provide a convex parameterization of the set of diagonalizable matrices with fixed eigenvalues, thereby allowing us to exploit the ideas of convexity theory. The fact that only diagonalizable matrices are represented in the grammian coordinates is not a cause for concern as it is a theoretical fact that if counterexamples to the rational dilation conjecture exist, then necessarily there are diagonalizable finite matrices that are counterexamples

2.2.1. Computationally Friendly Spectral Set and Dilation Conditions.

Here, we present two theorems which give conditions for a given $M \in \mathbb{C}^{n,n}$ to have Ω^- as a spectral set (Theorem 2.2.5) and for M to have a rational dilation to $\partial\Omega$ (Theorem 2.2.7). These conditions follow from two different generalizations to of the Herglotz Representation Theorem to the domain Ω (cf. Theorems 1.3.27 and 1.4.5 in Chapter 1). The representation theorems are formulated using certain canonical functions associated with the domain Ω, which we now briefly describe.

For u a continuous function on $\partial\Omega$, let $u^\wedge$ denote the solution to the Dirichlet problem with boundary data u. Let σ_r denote normalized arclength measure on ∂_r and set $\sigma = \sigma_0 + \cdots + \sigma_m$. Let $P_z(\lambda)$ denote the Poisson kernel with respect to σ, i.e. the unique 2-variable function defined for $z \in \Omega$ and $\lambda \in \partial\Omega$ such that

$$u^\wedge(z) = \int_{\partial\Omega} u(\lambda) P_z(\lambda) d\sigma(\lambda)$$

whenever u is a continuous function on $\partial\Omega$. (Note that we choose $d\sigma$ as our measure on $\partial\Omega$ rather than the more customary choice of arc length measure.) If $Rat(\Omega^-)$ denotes the rational functions with poles off Ω^-, then it is well-known that $\mathrm{Re}\,Rat(\Omega^-)|_{\partial\Omega}$ has codimension m in $L^2(d\sigma)$, the Hilbert space of real-valued, square integrable measurable functions on $\partial\Omega$ with inner product given by $\langle f, g \rangle = \int_{\partial\Omega} fg d\sigma$. Let $\varphi_1, \ldots, \varphi_m$ be an orthonormal basis for the orthogonal complement of $\mathrm{Re}\,Rat(\Omega^-)$ in $L^2(d\sigma)$.

We define two collections of holomorphic functions on Ω, the *Herglotz kernels* of the *first* and *second kinds*. In Theorem 2.2.5 below, the Herglotz kernel of the first kind will be used to give a concrete condition of a linear operator to have Ω^- as a spectral set. In Theorem 2.2.7 below, the Herglotz kernel of the second kind, will give a concrete condition for a linear operator to have a rational dilation to $\partial\Omega$.

Let T_Ω denote the $(m+1)$-dimensional torus defined by
$$T_\Omega = \partial_0 \times \ldots \partial_m = \{\alpha = (\alpha_r) : \alpha_r \in \partial_r \text{ for } r = 1, \ldots, m\}.$$
If $\alpha = (\alpha_0, \ldots, \alpha_m) \in T_\Omega$, define m vectors $\varphi_1(\alpha), \ldots, \varphi_m(\alpha) \in \mathbb{R}^{m+1}$ by the formula
$$\varphi_r(\alpha) = (\varphi_r(\alpha_0), \ldots, \varphi_r(\alpha_m)) \text{ for } r = 1, \ldots, m.$$
The following result from Chapter 1 provides the essential ingredient for defining Γ_α, the Herglotz kernel of the first kind.

PROPOSITION 2.2.1. *For each $\alpha \in T_\Omega$, there is a unique $(m+1)$-tuple of real numbers $\boldsymbol{w}(\alpha) = (w_0(\alpha), \ldots, w_m(\alpha))$ such that*

(2.2.2) $$\boldsymbol{w}(\alpha) \perp \varphi_r(\alpha) \text{ for } r = 1, \ldots, m \text{ and}$$

$$\sum_{r=0}^{m} w_r(\alpha) = 1$$

Furthermore, the entries of $\boldsymbol{w}(\alpha)$ are strictly positive.

Fix $z_0 \in \Omega$. Define for each $\alpha \in T_\Omega$ a holomorphic function Γ_α on Ω by requiring that

(2.2.3) $$Re\Gamma_\alpha(z) = \sum_{r=0}^{m} w_r(\alpha) P_z(\alpha_r), z \in \Omega, \text{ and}$$
$$Im\Gamma_\alpha(z_0) = 0.$$

Since Theorem 2.2.1 guarantees that $w_r(\alpha) > 0$ for each r, it follows that $\Gamma_\alpha(z)$ is an $m+1$ valent mapping from Ω to the right half plane with poles at $\alpha_0, \ldots, \alpha_m$.

Likewise, we now define H_λ, the Herglotz kernel of the second kind. For each fixed $\lambda \in \partial\Omega$, define a holomorphic function H_λ on Ω by requiring that

(2.2.4) $$ReH_\lambda(z) = P_z(\lambda) - \sum_{r=1}^{m} \varphi_r(\lambda)\varphi_r^\wedge(z), z \in \Omega, \text{ and}$$
$$ImH_\lambda(z_0) = 0.$$

These functions are discussed in detail in Chapter 1 and Chapter 3, and in particular Chapter 3 gives a complete description of the algorithms the authors developed to compute these functions.

For the goals of this memoir, the importance of the functions $\varphi_r, H_\lambda, \Gamma_\alpha$ is revealed by the following theorems.

THEOREM 2.2.5. *Let $M \in \mathbb{C}^{n,n}$, and assume that $\sigma(M) \subseteq \Omega$. M has Ω^- as a spectral set if and only if*

(2.2.6) $$Re\Gamma_\alpha(M) \geq 0 \text{ for all } \alpha \in T_\Omega.$$

THEOREM 2.2.7. *Let $M \in \mathbb{C}^{n,n}$, and assume that $\sigma(M) \subseteq \Omega$. M has a rational dilation to $\partial\Omega$ if and only if*

(2.2.8) *There exist self-adjoint matrices $\Phi_1, \ldots, \Phi_m$ such that*

$$ReH_\lambda(M) + \sum_{r=1}^{m} \varphi_r(\lambda)\Phi_r \geq 0 \text{ for all } \lambda \in \partial\Omega.$$

2.2.2. Functional Identities.

In this section we derive two functional identities involving the kernel functions Γ_α and H_λ defined in Section 2.2.1. The first functional identity, which expresses a Γ_α as a linear combination of H_λ's, allows a dramatic simplification of the rational dilation condition (2.2.8) of Theorem 2.2.7, in the case when the spectral set condition (2.2.6) of Theorem 2.2.5 holds extremally. The second identity, which gives a relationship between different Γ_α's, yields an equivalence relation on the set of Γ_α's that enables certain calculations to be significantly simplified.

The first functional identity, which relates the Herglotz kernels of the first and second kinds, is the content of the following proposition.

PROPOSITION 2.2.9. *If $\alpha = (\alpha_0, \alpha_1, \ldots, \alpha_m) \in T_\Omega$ then*

$$(2.2.10) \qquad \Gamma_\alpha = \sum_{r=0}^{m} w_r(\alpha) H_{\alpha_r}.$$

PROOF. Note that

$$Re\Gamma_\alpha(z) \stackrel{(2.2.3)}{=} \sum_{r=0}^{m} w_r(\alpha) P_z(\alpha_r)$$

$$\stackrel{(2.2.4)}{=} \sum_{r=0}^{m} w_r(\alpha) \left[ReH_{\alpha_r}(z) + \sum_{s=1}^{m} \varphi_s(\alpha_r) \varphi_s^{\wedge}(z) \right]$$

$$\stackrel{Prop.2.2.1}{=} \sum_{r=0}^{m} w_r(\alpha) ReH_{\alpha_r}(z).$$

The proposition now follows from the observation that the harmonic conjugate of $Re\Gamma_\alpha(z)$ is unique up to an imaginary constant, and the facts that $Im\Gamma_\alpha(z_0) = ImH_\lambda(z_0) = 0$ for all $\alpha \in T_\Omega$ and all $\lambda \in \partial\Omega$. □

Observe that since any $M \in \mathbb{C}^{n,n}$ that has a rational dilation to $\partial\Omega$ also has Ω^- as a spectral set, condition (2.2.8) of Theorem 2.2.7 must necessarily imply condition (2.2.6) of Theorem 2.2.5. Proposition 2.2.9 allows us to see this directly. Indeed, assume that M satisfies the hypotheses of Theorem 2.2.7 so that (2.2.8) holds. We see that for $\alpha \in T_\Omega$,

$$(2.2.11) \qquad Re\Gamma_\alpha(M) \stackrel{Prop.2.2.9}{=} \sum_{r=0}^{m} w_r(\alpha) ReH_{\alpha_r}(M),$$

$$(2.2.12) \qquad \stackrel{(2.2.2)}{=} \sum_{r=0}^{m} w_r(\alpha) \left(ReH_{\alpha_r}(M) + \sum_{s=1}^{m} \varphi_s(\alpha_r) \Phi_s \right),$$

$$(2.2.13) \qquad \stackrel{(2.2.8)}{\geq} 0.$$

Hence condition (2.2.6) holds.

Proposition 2.2.9 has other, more subtle, implications for the relationship between conditions (2.2.6) and (2.2.8). Indeed, if M is any matrix in $\mathbb{C}^{n,n}$ with that property that there exists $\alpha \in T_\Omega$ and a nonzero vector $u \in \mathbb{C}^n$ such that $Re\Gamma_\alpha(M)u = 0$, then Proposition 2.2.9 gives information about any matrix unknowns $\Phi_1, \ldots, \Phi_m$ that satisfy condition (2.2.8). Specifically, the following lemma obtains.

2.2. MATHEMATICAL PRELIMINARIES

LEMMA 2.2.14. *If $\alpha \in T_\Omega$, $u \in \mathbb{C}^n$, $u \neq 0$, and $Re\Gamma_\alpha(M)u = 0$ then the $m+1$ equations,*

$$\left(\sum_{s=1}^{m} \varphi_s(\alpha_r)\Phi_s\right) u = -ReH_{\alpha_r}(M)u, \text{ for } r = 0, \ldots, m,$$

hold whenever $\Phi_1, \ldots, \Phi_m$ satisfy (2.2.8).

PROOF. By (2.2.12), we have that

$$0 = Re\Gamma_\alpha(M)u = \left[\sum_{r=0}^{m} w_r(\alpha)\left(ReH_{\alpha_r}(M) + \sum_{s=1}^{m}\varphi_s(\alpha_r)\Phi_s\right)\right]u.$$

Since $\Phi_1, \ldots, \Phi_m$ satisfy (2.2.8), we have for each $r = 0, \ldots, m$,

$$\left(ReH_{\alpha_r}(M) + \sum_{s=1}^{m}\varphi_s(\alpha_r)\Phi_s\right) \geq 0.$$

The lemma now follows from Theorem 2.2.1 which guarantees that $w_r(\alpha) > 0$ for each $r = 0, \ldots, m$. □

Before continuing to the second functional identity, we remark that the hypothesis of Lemma 2.2.14 will be satisfied whenever M is a matrix satisfying condition (2.2.6) extremally. Thus, if a matrix M has Ω^- as a spectral set, condition (2.2.6) holds extremally, and one wants to check whether or not M has a rational dilation to $\partial\Omega$, then the lemma yields information about the matrix unknowns $\Phi_1, \ldots, \Phi_m$ of condition (2.2.8).

To elucidate the second functional identity, which relates the Γ_α's, we first note that the function Γ_α, an element of $Hol(\Omega)$, can be extended to a meromorphic function on a neighborhood of Ω. Let $\mathbb{H}$ denote the right half-plane in $\mathbb{C}$. The following result from Chapter 1 obtains.

PROPOSITION 2.2.15. *Let $f \in Hol(\Omega)$. The following are equivalent.*

(1) *There exist $\alpha \in T_\Omega, a > 0$ and $b \in \mathbb{R}$ such that $f = a\Gamma_\alpha + ib$.*
(2) *f extends meromorphically to a neighborhood $\mathcal{N}$ of Ω^- and, so extended, possesses the following properties:*

   f maps Ω into $\mathbb{H}$.

(2.2.16) *For all $r = 0, \ldots, m$, f has exactly one pole on ∂_r and*

   f is a 1-1 and onto map from ∂_r onto the extended imaginary axis

Since for $t \in \mathbb{R}$, $(z - it)^{-1}$ is an automorphism of $\mathbb{H}$, it follows from the above proposition that $(\Gamma_\alpha(z) - it)^{-1}$ is an affine scaling of Γ_β for some $\beta \in T_\Omega$. This observation leads to the following definition of an equivalence relation on the set of functions $\{\Gamma_\alpha : \alpha \in T_\Omega\}$.

DEFINITION 2.2.17. *For $\alpha \in T_\Omega, \beta \in T_\Omega$, $\Gamma_\alpha \sim \Gamma_\beta$ provided there exist $a > 0$ and $b, t \in \mathbb{R}$ such that*

(2.2.18) $$(\Gamma_\alpha - it)^{-1} = a\Gamma_\beta + ib.$$

This equivalence relation results in the following connection between the nullspaces of $Re\Gamma_\alpha(M)$ and $Re\Gamma_\beta(M)$.

LEMMA 2.2.19. *If* $\Gamma_\alpha \sim \Gamma_\beta$, *and* $M \in \mathbb{C}^{n,n}$ *then*

(2.2.20) $\quad\quad\quad\quad Re\Gamma_\alpha(M) \geq 0$ *if and only if* $Re\Gamma_\beta(M) \geq 0$.

and

(2.2.21) $\quad\quad\quad\quad\quad\quad u \in ker\ Re\Gamma_\alpha(M)$ *if and only if*
$$[(\Gamma_\alpha - it)(M)]\,u \in ker Re\Gamma_\beta(M) \text{ for all } t \in \mathbb{R}.$$

PROOF. By the equivalence relation of Definition 2.2.17, we have

(2.2.22) $\quad\quad aRe\Gamma_\beta(M) = Re\left[(\Gamma_\alpha - it)^{-1}(M)\right]$

(2.2.23) $\quad\quad\quad\quad\quad\quad = \dfrac{1}{2}\left[(\Gamma_\alpha - it)^{-1}(M) + (\Gamma_\alpha - it)^{-1}(M)^*\right]$

(2.2.24) $\quad\quad\quad\quad\quad\quad = (\Gamma_\alpha - it)^{-1}(M)^* \left[Re\Gamma_\alpha(M)\right](\Gamma_\alpha - it)^{-1}(M).$

This formula implies the lemma. $\quad\square$

We will find it convenient to replace the parameter it in equation (2.2.18) by the quantity $\Gamma_\alpha(\lambda)$ for suitable $\lambda \in \partial\Omega$. Such a substitution is possible by Proposition 2.2.15, which guarantees that for each $t \in \mathbb{R}$ the equation

$$it = \Gamma_\alpha(\lambda)$$

has exactly $m+1$ solutions, one solution from each component of $\partial\Omega$. Furthermore, if for fixed $\alpha \in T_\Omega$, and for each $\lambda \in \partial\Omega$ we define $\alpha(\lambda)$ by requiring that

(2.2.25) $\quad\quad\quad\quad\quad\quad\quad\quad \alpha_r(\lambda) \in \partial_r$

and

(2.2.26) $\quad\quad\quad\quad\quad\quad \Gamma_\alpha(\alpha_r(\lambda)) = \Gamma_\alpha(\lambda),$

then $\Gamma_\alpha \sim \Gamma_{\alpha(\lambda)}$. In particular, note that if $\lambda \in \partial_r$, then $\alpha(\lambda)_r = \lambda$. Hence, Lemma 2.2.19 can be reformulated in the following way.

LEMMA 2.2.27. *Let* $M \in \mathbb{C}^{n,n}$ *and let* $u \in \mathbb{C}^n$ *with* $u \neq 0$.

(i): $Re\Gamma_\alpha(M) \geq 0$ *if and only if* $Re\Gamma_{\alpha(\lambda)}(M) \geq 0$ *for all* $\lambda \in \partial\Omega$.
(ii): $u \in ker Re\Gamma_\alpha(M)$ *if and only if*

$$(\Gamma_\alpha - \Gamma_\alpha(\lambda))(M)u \in ker Re\Gamma_{\alpha(\lambda)}(M) \text{ for all } \lambda \in \partial\Omega.$$

The equivalence relation introduced in Definition 2.2.17 yields two significant dividends. The first of these, Proposition 2.2.28 below is a dramatic strengthening of Lemma 2.2.14 which will allow us in Section 2.3 to associate a linear system to any matrix M that satisfies the condition (2.2.6) sufficiently extremally. This linear systems determines any solution $\Phi_1, \Phi_2, \ldots, \Phi_m$ of (2.2.8) uniquely. The second dividend, Proposition 2.2.32 below, is a simplification of condition (2.2.6) that is important both theoretically (cf. the notion of extremality in Section 2.3) and computationally.

The first dividend is an immediate consequence of Lemmas 2.2.14 and 2.2.27.

PROPOSITION 2.2.28. *If $\alpha \in T_\Omega$, $u \in \mathbb{C}^n$, $u \neq 0$, and $\mathrm{Re}\,\Gamma_\alpha(M)u = 0$ then the equations,*

$$(2.2.29) \quad \left[\sum_{s=1}^{m} \varphi_s(\alpha_r(\lambda))\Phi_s\right] (\Gamma_\alpha - \Gamma_\alpha(\lambda))(M)u$$

$$= -\mathrm{Re}\,H_{\alpha_r(\lambda)}(M)(\Gamma_\alpha - \Gamma_\alpha(\lambda))(M)u$$

hold for $r = 0, \ldots, m$ and for all $\lambda \in \partial\Omega$, whenever $\Phi_1, \ldots, \Phi_m$ satisfy (2.2.8).

At first glance, this proposition seems to yield too much information about the unknowns $\Phi_1, \ldots, \Phi_m$. However, we can make several observations. First, although for each fixed $\lambda \in \partial\Omega$, there are $m + 1$ equations obtained by setting $r = 0, \ldots, m$, in fact only m of these equations are linearly independent. Second, there is an equation (2.2.29) for each $\lambda \in \partial\Omega$ ($m + 1$ parameters), but the system depends on $\Gamma_\alpha(\alpha(\lambda))$ (1 parameter). Third, by varying $\lambda \in \partial\Omega$ one obtains an infinite number of equations. Through the computations described in the second chapter of this memoir, we observe empirically that these equations are consistent, although the authors have not been able to prove or explain this observation. Finally, the span of $\{u_\lambda\}_{\lambda \in \partial\Omega}$ is a two-dimensional subspace of $\mathbb{C}^n$. Hence, even with infinite number of $\lambda \in \partial\Omega$, the unknowns $\Phi_1, \ldots, \Phi_m$ are, at best determined only on a two-dimensional subspace.

The second dividend of the equivalence relation consists of the simplification of condition (2.2.6). By exploiting the equivalence relation of Definition 2.2.17, we reduce the condition from a system of matrix inequalities continuously parameterized by the $(m + 1)$-dimensional torus to a system continuously parameterized by the m-dimensional torus. First, we make a preliminary observation about the equivalence relation given in Definition 2.2.17.

LEMMA 2.2.30. *If $\alpha \in T_\Omega$ and $\lambda \in \partial_r$, then there exists $\beta \in T_\Omega$ such that $\Gamma_\beta \sim \Gamma_\alpha$ and $\beta_r = \lambda$.*

PROOF. Note that the lemma trivially holds if $\lambda = \alpha_r$. So assume $\lambda \neq \alpha_r$. Since Γ_α maps $\partial_r \setminus \{\alpha_r\}$ bijectively onto $i\mathbb{R}$, there exists a unique $t \in \mathbb{R}$ such that $\Gamma_\alpha(\lambda) = it$. Now by Proposition 2.2.15 there exists $\beta \in T_\Omega, a > 0$, and $b \in \mathbb{R}$ such that $(\Gamma_\alpha(z) - it)^{-1} = a\Gamma_\beta + ib$. Since Γ_β has a single pole on each ∂_r located at β_r, and $(\Gamma_\alpha(z) - it)^{-1}$ has a pole at $\lambda \in \partial_r$, it must be the case that $\beta_r = \lambda$. □

Evidently, Lemma 2.2.30 implies that if for a fixed $\lambda \in \partial\Omega$, we define

$$(2.2.31) \quad T_\Omega(\lambda) = \{\alpha \in T_\Omega : \alpha_r = \lambda\},$$

then for any $\alpha \in T_\Omega$, there exists $\beta \in T_\Omega(\lambda)$ such that $\Gamma_\beta \sim \Gamma_\alpha$. Thus, Theorem 2.2.5 and Lemma 2.2.30 imply the following result.

PROPOSITION 2.2.32. *Fix $\lambda \in \partial\Omega$. Let $M \in \mathbb{C}^{n,n}$, and assume that $\sigma(M) \subseteq \Omega$. M has Ω^- as a spectral set if and only if*

$$(2.2.33) \quad \mathrm{Re}\,\Gamma_\alpha(M) \geq 0 \text{ for all } \alpha \in T_\Omega(\lambda).$$

2.2.3. Grammian Coordinates, the Spectral Set Body, and the Dilation Body.

We now introduce grammian coordinates, which express a diagonalizable matrix in terms of the grammian of its eigenvectors. Let $M \in \mathbb{C}^{n,n}$ and suppose $\sigma(M)$ consists of the n distinct eigenvalues $z_1, \ldots, z_n \in \mathbb{C}$ with corresponding eigenvectors $u_1, \ldots, u_n$. It is easy to see that if G denotes the grammian of the vectors

$u_1, \ldots, u_n$ (i.e. the $n \times n$ positive definite matrix $G = [\langle u_j, u_i \rangle]$) and D_z denotes the diagonal matrix whose ii^{th} entry is z_i, then M is unitarily equivalent to the matrix $G^{\frac{1}{2}} D_z G^{-\frac{1}{2}}$. Conversely, if $G > 0$ and M is defined by $M = G^{\frac{1}{2}} D_z G^{-\frac{1}{2}}$, then M is an $n \times n$ matrix with $\sigma(M) = \{z_1, \ldots, z_n\}$. Notice that if u is an eigenvector for M, then so also is cu when $c \neq 0$. Thus, G depends upon the scaling of the eigenvectors of M. In this memoir, we say that G is *strongly normalized* if $g_{ii} = \|u_i\|^2 = 1$ for each $i = 1, 2, \ldots, n$. In Section 2.5.2, we shall also require a different normalization: we say that G is *weakly normalized* if $\text{tr} G = n$ (i.e. $\sum_{i=1}^{n} \|u_i\|^2 = n$).

Fix n distinct points $z_1, \ldots, z_n$ in Ω. We define the following sets of strongly normalized grammians:

(2.2.34)
$$\mathcal{S}_z = \{G \in \mathbb{C}^{n,n} : G > 0, g_{ii} = 1, \text{ and } G^{\frac{1}{2}} D_z G^{-\frac{1}{2}} \text{ has } \Omega^- \text{ as a spectral set}\},$$

(2.2.35)
$$\mathcal{D}_z = \{G \in \mathbb{C}^{n,n} : G > 0, g_{ii} = 1, \text{ and } G^{\frac{1}{2}} D_z G^{-\frac{1}{2}} \text{ has a rational dilation to } \partial \Omega\}.$$

We call $\mathcal{S}_z$ the *spectral set body*, and we call $\mathcal{D}_z$ *dilation body*. Evidently, with these notations, it is tautological that $\mathcal{D}_z \subseteq \mathcal{S}_z$. Furthermore, if the dilation conjecture is true then $\mathcal{D}_z = \mathcal{S}_z$. To probe the relationship between the spectral set bodies and the dilation bodies, we first reformulate Theorems 2.2.5 and 2.2.7 in grammian coordinates.

Specifically, if $M \in \mathbb{C}^{n,n}$, and M is unitarily equivalent to $G^{\frac{1}{2}} D_z G^{-\frac{1}{2}}$, then Theorem 2.2.5 implies that M has Ω^- as a spectral set if and only if

$$\text{Re} \Gamma_\alpha (G^{\frac{1}{2}} D_z G^{-\frac{1}{2}}) \geq 0 \text{ for all } \alpha \in T_\Omega.$$

We see that

(2.2.36) $$\text{Re} \Gamma_\alpha (G^{\frac{1}{2}} D_z G^{-\frac{1}{2}}) = G^{-\frac{1}{2}} \left[g_{ij} \left(\tfrac{1}{2} (\overline{\Gamma_\alpha(z_i)} + \Gamma_\alpha(z_j)) \right) \right] G^{-\frac{1}{2}}.$$

Thus, if we define an $n \times n$ matrix $\gamma(\alpha) = [\gamma_{ij}(\alpha)]_{i,j=1}^{n}$ by setting

(2.2.37) $$\gamma_{ij}(\alpha) = \tfrac{1}{2} \left(\overline{\Gamma_\alpha(z_i)} + \Gamma_\alpha(z_j) \right),$$

and we denote the Schur product of two matrices A and B by $A * B$, we obtain the following description of $\mathcal{S}_z$.

PROPOSITION 2.2.38. *$G \in \mathcal{S}_z$ if and only if $G > 0$, G is strongly normalized and*

(2.2.39) $$\gamma(\alpha) * G \geq 0 \text{ for all } \alpha \in T_\Omega.$$

Likewise, we can use Theorem 2.2.7 to derive the following concrete description of $\mathcal{D}_z$, where for $\lambda \in \partial \Omega$ we define $h(\lambda)$ to be the $n \times n$ matrix whose ij^{th} entry is

$$h_{ij}(\lambda) = \overline{H_\lambda(z_i)} + H_\lambda(z_j).$$

PROPOSITION 2.2.40. *$G \in \mathcal{D}_z$ if and only if $G > 0$, G is strongly normalized and there exist m self-adjoint $n \times n$ matrices $X_1, \ldots, X_m$ such that*

(2.2.41) $$h(\lambda) * G + \sum_{r=1}^{m} \varphi_r(\lambda) X_r \geq 0 \text{ for all } \lambda \in \partial \Omega.$$

2.2. MATHEMATICAL PRELIMINARIES

Propositions 2.2.38 and 2.2.40 yield an effective tools for analyzing the spectral set and dilation bodies, and in particular for probing the relationship between the bodies via machine computation. Some important consequences of the propositions are the following facts about the geometry of the spectral set and dilation bodies.

PROPOSITION 2.2.42. $\mathcal{S}_z$ and $\mathcal{D}_z$ are convex and compact.

PROOF. Convexity following immediately from conditions (2.2.39) and (2.2.41). See Proposition 1.7.7 in Chapter 1 for the proof of compactness. (Note that although the normalization conditions used to define $\mathcal{S}_z$ and $\mathcal{D}_z$ in Section 2.2.2 and $\mathcal{S}_z(\Omega)$ and $\mathcal{D}_z(\Omega)$ in Chapter 1, Section 1.7 are slightly different, the proofs of compactness are identical.)

Alternatively, the compactness of $\mathcal{S}_z$ can be deduced from the compactness of $\mathcal{S}_z(\Omega)$ by noting that $G \in \mathcal{S}_z$ if and only if G is of the form

$$G = D^*_{(1,e^{i\theta_2},\ldots,e^{i\theta_n})} G_1 D_{(1,e^{i\theta_2},\ldots,e^{i\theta_n})},$$

where $G_1 \in \mathcal{S}_z(\Omega)$. Thus, $\mathcal{S}_z$ is the continuous image of the compact set $[0, 2\pi]^{n-1} \times \mathcal{S}_z(\Omega)$. Similarly, $\mathcal{D}_z$ is the continuous image of the compact set $[0, 2\pi]^{n-1} \times \mathcal{D}_z(\Omega)$. □

We now translate several key results from Section 2.2.2 into grammian coordinates. The next proposition follows immediately from (2.2.2) and (2.2.10).

PROPOSITION 2.2.43. If $G \in \mathbb{C}^{n,n}$ and $G > 0$, then for $Y_1, \ldots, Y_m \in \mathbb{C}^{n,n}$

$$(2.2.44) \qquad \gamma(\alpha) * G = \sum_{r=0}^{m} w_r(\alpha) \left[h(\alpha_r) * G + \sum_{s=1}^{m} \varphi_s(\alpha_r) Y_s \right]$$

Proposition 2.2.46 below is a reformulation of Lemma 2.2.27 and follows from the following computational lemma.

LEMMA 2.2.45. Let $\alpha \in T_\Omega$, let $\alpha(\lambda)$ be defined as in (2.2.25) and (2.2.26), and let $D_{\Gamma_\alpha(z)}$ denote the $n \times n$ diagonal matrix whose ii^{th} entry is $\Gamma_\alpha(z_i)$. If $G > 0$, then for each $\lambda \in \partial\Omega$ there exists $a_{\alpha(\lambda)} > 0$ such that

$$a_{\alpha(\lambda)} \left[\gamma(\alpha(\lambda)) * G \right] = \left[(D_{\Gamma_\alpha(z)} - \Gamma_\alpha(\lambda))^{-1} \right]^* [\gamma(\alpha) * G] \left[D_{\Gamma_\alpha(z)} - \Gamma_\alpha(\lambda) \right]^{-1}.$$

PROOF. Let $M = G^{\frac{1}{2}} D_z G^{-\frac{1}{2}}$. By (2.2.24), and the definition of $\alpha(\lambda)$, there exists an $a_{\alpha(\lambda)}$ such that

$$a_{\alpha(\lambda)} Re\Gamma_{\alpha(\lambda)}(M) = (\Gamma_\alpha - \Gamma_\alpha(\lambda))^{-1} (M)^* [Re\Gamma_\alpha(M)] (\Gamma_\alpha - \Gamma_\alpha(\lambda))^{-1} (M).$$

From (2.2.36) and (2.2.37), we obtain

$$a_{\alpha(\lambda)} G^{-\frac{1}{2}} [\gamma(\alpha(\lambda)) * G] G^{-\frac{1}{2}}$$
$$= (\Gamma_\alpha - \Gamma_\alpha(\lambda))^{-1} (M)^* \left[G^{-\frac{1}{2}} (\gamma(\alpha) * G) G^{-\frac{1}{2}} \right] (\Gamma_\alpha - \Gamma_\alpha(\lambda))^{-1} (M).$$

Hence,

$$a_{\alpha(\lambda)}(\gamma(\alpha(\lambda)) * G)$$
$$= \left[G^{\frac{1}{2}}\left((\Gamma_\alpha(M) - \Gamma_\alpha(\lambda))^{-1}\right)^* G^{-\frac{1}{2}}\right][\gamma(\alpha) * G]\left[G^{-\frac{1}{2}}(\Gamma_\alpha(M) - \Gamma_\alpha(\lambda))^{-1} G^{\frac{1}{2}}\right]$$
$$= \left[(D_{\Gamma_\alpha(z)} - \Gamma_\alpha(\lambda))^{-1}\right]^* [\gamma(\alpha) * G]\left[D_{\Gamma_\alpha(z)} - \Gamma_\alpha(\lambda)\right]^{-1}.$$

□

PROPOSITION 2.2.46. *Let $G \in \mathbb{C}^{n,n}$ with $G > 0$, and let $u \in \mathbb{C}^n$ with $u \neq 0$.*
(1) *$\gamma(\alpha) * G \geq 0$ if and only if*
$$\gamma(\alpha(\lambda)) * G \geq 0 \text{ for all } \lambda \in \partial\Omega.$$
(2) *$u \in \ker(\gamma(\alpha) * G)$ if and only if*
$$\left[D_{\Gamma_\alpha(z)} - \Gamma_\alpha(\lambda)\right] u \in \ker(\gamma(\alpha(\lambda)) * G) \text{ for all } \lambda \in \partial\Omega.$$

Finally, we formulate Propositions 2.2.28 and Proposition 2.2.32 from Section 2.2.2 in grammian coordinates.

PROPOSITION 2.2.47. *Let $\alpha \in T_\Omega$, and let $u \in \mathbb{C}^n$ with $u \neq 0$. For $\lambda \in \partial\Omega$ define $u_\lambda \in \mathbb{C}^n$ by*
$$u_\lambda = \left[D_{\Gamma_\alpha(z)} - \Gamma_\alpha(\lambda)\right] u.$$
If $(\gamma(\alpha) * G) u = 0$, then the equations

$$(2.2.48) \qquad \left(\sum_{s=1}^m \varphi_s(\alpha_r(\lambda))X_s\right)u_\lambda = -(h(\alpha_r(\lambda)) * G)u_\lambda$$

hold for $r = 0, \ldots, m$ and for all $\lambda \in \partial\Omega$, whenever $X_1, \ldots, X_m$ satisfy (2.2.41).

PROPOSITION 2.2.49. *Fix $\lambda \in \partial\Omega$. $G \in \mathcal{S}_z$ if and only if $G > 0$, G is strongly normalized, and*

$$(2.2.50) \qquad G * \gamma(\alpha) \geq 0 \text{ for all } \alpha \in T_\Omega(\lambda).$$

2.3. Analysis of the Dilation Condition for Nonsingularly Hyperextremal Grammians

In this section, we make precise the notion that some grammians meet the spectral set condition (2.2.50) more sharply than others. We define *extremal* grammians (Definition 2.3.2) and the stronger notion of *hyperextremal* grammians (Definition 2.3.5). In grammian coordinates, the rational dilation conjecture is equivalent to the statement that $\mathcal{D}_z$, which *a priori* is a compact convex subset of $\mathcal{S}_z$, is in fact equal to $\mathcal{S}_z$. Thus, to probe the rational dilation conjecture, it makes sense to determine if the extremals of Definition 2.3.2 – which are exactly the elements of $\partial \mathcal{S}_z$ – are in $\mathcal{D}_z$. Furthermore, it turns out that the hyperextremals of Definition 2.3.5 are a subset of $\partial \mathcal{S}_z$ containing the extreme points of $\mathcal{S}_z$. Hence, by convexity

$$\mathcal{D}_z = \mathcal{S}_z,$$

if and only if

$$\text{extremals} \subseteq \mathcal{D}_z,$$

2.3. DILATION CONDITION FOR NONSINGULARLY HYPEREXTREMAL GRAMMIANS

if and only if

$$\text{hyperextremals} \subseteq \mathcal{D}_z.$$

In particular, the rational dilation conjecture is equivalent to the assertion that the hyperextremal grammians all have rational dilations.

If $m = 2$ and $n = 4$, then generically the question of whether or not a hyperextremal grammian is in $\mathcal{D}_z$ reduces to the simpler question of the consistency of a linear system. Specifically, we introduce the notion of a *nonsingular hyperextremal grammian* Ξ (Definition 2.3.17), and see that the associated linear system uniquely determines the matrices X_1 and X_2 of Proposition 2.2.40. Thus, to determine whether or not $\Xi \in \mathcal{D}_z$, we simply check whether or not this system is consistent and if so, check the inequality (2.2.41) with the computed matrices X_1 and X_2. Since the matrices X_1 and X_2 of Proposition 2.2.40 are self-adjoint, we obtain two lower bounds for the distance between Ξ and $\mathcal{D}_z$ in terms of two different measures of the non-self-adjointness of the computed X_1 and X_2 (Propositions 2.3.55 and 2.3.59).

We now turn to the first definition. Fix a point $\lambda^* \in \partial_0$, the outer component of $\partial \Omega$, and denote $T_\Omega(\lambda^*)$, the m-torus defined in (2.2.31) by T_Ω^0. From Proposition 2.2.49 it is easy to see that if $G > 0$ is strongly normalized, then $G \in \partial \mathcal{S}_z$ if and only if

$$\gamma(\alpha) * G \geq 0 \text{ for all } \alpha \in T_\Omega^0$$

and

(2.3.1) There exists $\alpha \in T_\Omega^0$ and $u^* \in \mathbb{C}^n$ with $\|u^*\| = 1$ such that

$$(\gamma(\alpha^*) * G) u^* = 0.$$

Thus, $G \in \partial \mathcal{S}_z$ if and only if (2.2.50) holds extremally. Accordingly, we make the following definition.

DEFINITION 2.3.2. *If $G > 0$ is strongly normalized, we say that G is **extremal** provided*

(2.3.3) $$\gamma(\alpha) * G \geq 0 \text{ for all } \alpha \in T_\Omega^0$$

and

(2.3.4) *There exists $\alpha \in T_\Omega^0$ and $u^* \in \mathbb{C}^n$ with $\|u^*\| = 1$ such that*
$$(\gamma(\alpha^*) * G) u^* = 0.$$

A stronger notion of extremality for $\mathcal{S}_z$ is recorded in the following definition.

DEFINITION 2.3.5. *If $G > 0$ is strongly normalized, we say that G is **hyperextremal** provided*

(2.3.6) $$\gamma(\alpha) * G \geq 0 \text{ for all } \alpha \in T_\Omega^0$$

and

(2.3.7) *There exists $\alpha^*, \beta^* \in T_\Omega^0$ with $\alpha^* \neq \beta^*$ and $u^*, v^* \in \mathbb{C}^n$ with*
$$\|u^*\| = \|v^*\| = 1 \text{ such that } [\gamma(\alpha^*) * G] u^* = 0, \text{ and}$$
$$[\gamma(\beta^*) * G] v^* = 0.$$

TABLE 2.1. Critical 4-tuples.

l	π_l	ζ_l	y_l	r_l
1	α^*	α_1^*	u^*	1
2	α^*	α_2^*	u^*	2
3	$\alpha^*(\lambda_1)$	λ_1	$\frac{u_{\lambda_1}}{\|u_{\lambda_1}\|}$	1
4	$\alpha^*(\lambda_2)$	λ_2	$\frac{u_{\lambda_2}}{\|u_{\lambda_2}\|}$	2
5	β^*	β_1^*	v^*	1
6	β^*	β_2^*	v^*	2
7	$\beta^*(\lambda_1)$	λ_1	$\frac{v_{\lambda_1}}{\|v_{\lambda_1}\|}$	1
8	$\beta^*(\lambda_2)$	λ_2	$\frac{v_{\lambda_2}}{\|v_{\lambda_2}\|}$	2

For the remainder of this section, we assume that $m = 2$ and $n = 4$. Accordingly, we fix a two-holed domain Ω, choose spectral points $z = (z_1, z_2, z_3, z_4) \subseteq \Omega$, and assume that Ξ is a hyperextremal grammian in $\mathcal{S}_z$, with $\alpha^*, \beta^* \in T_\Omega^0$ and $u^*, v^* \in \mathbb{C}^4$ as in Definition 2.3.5, and choose $\lambda_1 \in \partial_1$ and $\lambda_2 \in \partial_2$. These objects: $m, n, \Omega, z, \Xi, \lambda_1$ and λ_2 are fixed throughout the remainder of this section.

Observe that the parameters $\alpha \in T_\Omega^0$, $\lambda \in \partial\Omega$, $u \in \mathbb{C}^n$ and $r \in \{0, 1, 2\}$ appear in the statement of Proposition 2.2.47. For each integer l, $1 \leq l \leq 8$, we obtain a linear equation in X_1 and X_2 by applying Proposition 2.2.47 with

$$(\alpha, \lambda, u, r) = (\pi_l, \zeta_l, y_l, r_l)$$

where $(\pi_l, \zeta_l, y_l, r_l)$ is given in Table 2.1. Note that

(2.3.8) $$\zeta_l = (\pi_l)_{r_l}.$$

Also observe that Proposition 2.2.46 implies that

(2.3.9) $$[\gamma(\pi_l) * \Xi]\, y_l = 0 \text{ for each } l.$$

Finally, observe that since u^* and v^* are assumed to be unit vectors in Definition 2.3.5 that

(2.3.10) $$\|y_l\| = 1 \text{ for each } l.$$

If X_1 and X_2 satisfy (2.2.41) with $G = \Xi$, i.e.

(2.3.11) $$h(\lambda) * \Xi + \varphi_1(\lambda)X_1 + \varphi_2(\lambda)X_2 \geq 0 \text{ for all } \lambda \in \partial\Omega,$$

then for each $l = 1, \ldots, 8$, we apply Proposition 2.2.47 with $G = \Xi$ and $(\alpha, \lambda, u, r) = (\pi_l, \zeta_l, y_l, r_l)$ to obtain an equation (2.2.48) which we write in the form

(2.3.12) $$[X_1 X_2] \begin{pmatrix} \varphi_1(\zeta_l) y_l \\ \varphi_2(\zeta_l) y_l \end{pmatrix} = -(h(\zeta_l) * \Xi) y_l.$$

We express these eight vector equations as a linear system in the following way. Let $A_{\Xi, \lambda_1, \lambda_2}$ be the 8×8 matrix whose l^{th} column, a_l, is given by

(2.3.13) $$a_l = \begin{pmatrix} \varphi_1(\zeta_l) y_l \\ \varphi_2(\zeta_l) y_l \end{pmatrix}.$$

Let $B_{\Xi, \lambda_1, \lambda_2}$ be the 4×8 matrix whose l^{th} column, b_l, is given by

(2.3.14) $$b_l = -(h(\zeta_l) * \Xi) y_l,$$

2.3. DILATION CONDITION FOR NONSINGULARLY HYPEREXTREMAL GRAMMIANS

and let $\mathbf{X} = [X_1 X_2]$ be the 4×8 matrix consisting of the 4×4 blocks X_1 and X_2. Clearly, with this notation, the equations (2.3.12) for $l = 1, \ldots, 8$, can be written as
$$\mathbf{X} A_{\Xi, \lambda_1, \lambda_2} = B_{\Xi, \lambda_1, \lambda_2}.$$
The following proposition formalizes the above discussion.

PROPOSITION 2.3.15. *If Ξ is hyperextremal, $\lambda_1, \lambda_2 \in \partial\Omega$, and if X_1 and X_2 are $n \times n$ matrices that satisfy (2.3.11), then $\mathbf{X} = [X_1 X_2]$ solves the equation*

(2.3.16) $$\mathbf{X} A_{\Xi, \lambda_1, \lambda_2} = B_{\Xi, \lambda_1, \lambda_2}.$$

For a hyperextremal grammian Ξ, the linear system $\mathbf{X} A_{\Xi, \lambda_1, \lambda_2} = B_{\Xi, \lambda_1, \lambda_2}$ gives exactly thirty-two complex equations in the thirty-two complex unknowns that form the entries of X_1 and X_2. The following definition formalizes the case when this system is nonsingular.

DEFINITION 2.3.17. *If Ξ is hyperextremal, we say that Ξ is **nonsingularly hyperextremal** provided there exist $\lambda_1, \lambda_2 \in \partial\Omega$ such that the linear system*
$$\mathbf{X} A_{\Xi, \lambda_1, \lambda_2} = B_{\Xi, \lambda_1, \lambda_2}$$
is nonsingular.

In the case when Ξ is nonsingularly hyperextremal, Proposition 2.3.15 allows us to solve (2.2.41), the feasibility of a complicated linear matrix inequality, by solving (2.3.16), a relatively trivial linear system. In particular, we obtain the following result.

PROPOSITION 2.3.18. *Let Ξ be nonsingularly hyperextremal, let $\lambda_1 \in \partial_1, \lambda_2 \in \partial_2$, and assume that $\mathbf{X} = [X_1 X_2]$ solves the linear system $\mathbf{X} A_{\Xi, \lambda_1, \lambda_2} = B_{\Xi, \lambda_1, \lambda_2}$. If $\Xi \in \mathcal{D}_z$, then X_1 and X_2 are self-adjoint.*

PROOF. Since $\Xi \in \mathcal{D}_z$, by Proposition 2.2.40 there exist self-adjoint 4×4 matrices Y_1 and Y_2 such that
$$h(\lambda) * \Xi + \varphi_1(\lambda) Y_1 + \varphi_2(\lambda) Y_2 \geq 0 \text{ for all } \lambda \in \partial\Omega.$$
Consequently, setting $\mathbf{Y} = [Y_1 Y_2]$, we obtain by Proposition 2.3.15,
$$Y A_{\Xi, \lambda_1, \lambda_2} = B_{\Xi, \lambda_1, \lambda_2}.$$
Since Ξ is nonsingularly hyperextremal, it follows that $\mathbf{X} = \mathbf{Y}$, and hence X_1 and X_2 are self-adjoint. □

In the computational experiments that we describe in Section 2.6, we generate – via a procedure outlined in Section 2.5.2 – a number of grammians Ξ that are "approximately" nonsingular hyperextremal grammians. (A notion that we make precise later.) However, for these grammians the solution $\mathbf{X} = [X_1 X_2]$ of the linear system (2.3.16) does not yield self-adjoint matrices X_1 and X_2, a state of affairs that is inconsistent with Proposition 2.3.18. Accordingly, we speculate that such grammians are counterexamples to the rational dilation conjecture. However, in order to justify this speculation, we need to develop a robust version of Proposition 2.3.18.

To this end, note that Proposition 2.3.18 states that if $\Xi \in \mathcal{D}_z$ then X_1 and X_2 are self-adjoint. We wish to strengthen this result to say that if Ξ is "close" to $\mathcal{D}_z$ then the matrices X_1 and X_2 obtained by solving the linear system (2.3.16) are "close" to being self-adjoint. Thus, we wish to control the non-selfadjointness of

X_1 and X_2 by the distance to the dilation body. Proposition 2.3.41 below provides a result of this form.

In order to state the proposition rigorously, we must make precise the measures of non-self-adjointness and the distance to the dilation body. For $M \in \mathbb{C}^{n,n}$, let $\|M\|$ denote the operator norm, and let $\|M\|_2 = \left(\sum_{i,j=1}^n |m_{ij}|^2\right)^{\frac{1}{2}}$ denote the Hilbert-Schmidt norm. It is well known that if $M, N \in \mathbb{C}^{n,n}$, then

$$\|MN\|_2 \leq \|M\| \|N\|_2. \tag{2.3.19}$$

For each $l = 1, \ldots, 8$, define two constants:

$$C_l^\gamma = \max_i |\Gamma_{\pi_l}(z_i)| \tag{2.3.20}$$

$$C_l^h = \max_i |H_{\zeta_l}(z_i)|. \tag{2.3.21}$$

Further, define for each $l = 1, \ldots, 8$, the error function

$$E_l(\epsilon) = \frac{C_l^\gamma}{w_{r_l}(\pi_l)} \left(\sqrt{\|\Xi\|} + \epsilon\right) \epsilon^{\frac{1}{2}} + C_l^h \epsilon.$$

Note that for ϵ small, $E_l(\epsilon) \approx C\epsilon^{\frac{1}{2}}$ where C is constant. The following lemma asserts that if a nonsingularly hyperextremal Ξ is close to an element $G \in \mathcal{D}_z$ and $\mathbf{Y} = [Y_1 Y_2]$ solves condition (2.2.41), then $\mathbf{Y}$ approximately solves the system $\mathbf{X} A_{\Xi, \lambda_1, \lambda_2} = B_{\Xi, \lambda_1, \lambda_2}$.

LEMMA 2.3.22. *Assume that Ξ is nonsingularly hyperextremal and that there exists $G \in \mathcal{D}_z$ such that $\|G - \Xi\| < \epsilon$. If Y_1 and Y_2 are 4×4 matrices that satisfy*

$$h(\lambda) * G + \varphi_1(\lambda) Y_1 + \varphi_2(\lambda) Y_2 \geq 0 \text{ for all } \lambda \in \partial\Omega, \tag{2.3.23}$$

and $\mathbf{Y} = [Y_1 Y_2]$, then

$$\mathbf{Y} A_{\Xi, \lambda_1, \lambda_2} = B_{\Xi, \lambda_1, \lambda_2} + \delta, \tag{2.3.24}$$

where δ is a 4×8 matrix whose l^{th} column, δ_l, satisfies $\|\delta_l\|_2 \leq E_l(\epsilon)$.

PROOF. For each $l = 1, \ldots, 8$ and for $r = 0, 1, 2$, we have by (2.3.23) that

$$h((\pi_l)_r) * G + \varphi_1((\pi_l)_r) Y_1 + \varphi_2((\pi_l)_r) Y_2 \geq 0 \tag{2.3.25}$$

Recall from Proposition 2.2.1 that $w_r(\alpha) > 0$ for all $\alpha \in T_\Omega$, and recall that from (2.3.8) that $(\pi_l)_{r_l} = \zeta_l$. Hence,

$$w_{r_l}(\pi_l) \left[h(\zeta_l) * G + \varphi_1(\zeta_l) Y_1 + \varphi_2(\zeta_l) Y_2\right] \tag{2.3.26}$$
$$\leq \sum_{r=0}^{2} w_r(\pi_l) \left[h(\pi_l)_r * G + \varphi_1((\pi_l)_r) Y_1 + \varphi_2((\pi_l)_r) Y_2\right].$$

By Proposition 2.2.43 we have

$$\gamma(\pi_l) * G = \sum_{r=0}^{2} w_r(\pi_l) \left[h(\pi_l)_r * G + \varphi_1((\pi_l)_r) Y_1 + \varphi_2((\pi_l)_r) Y_2\right]. \tag{2.3.27}$$

Let $A_l = h(\zeta_l) * G + \varphi_1(\zeta_l) Y_1 + \varphi_2(\zeta_l) Y_2$. Combining (2.3.26) and (2.3.27) we see that

$$w_{r_l}(\zeta_l) A_l \leq \gamma(\pi_l) * G. \tag{2.3.28}$$

2.3. DILATION CONDITION FOR NONSINGULARLY HYPEREXTREMAL GRAMMIANS

Now, we compute

(2.3.29) $$\|\gamma(\pi_l) * G\| = \left\|\frac{1}{2}\left(D^*_{\Gamma_{\pi_l}(z)}G + GD_{\Gamma_{\pi_l}(z)}\right)\right\|$$

(2.3.30) $$\leq \|D_{\Gamma_{\pi_l}(z)}\|\|G\|$$

(2.3.31) $$\leq C^\gamma_l \|G\|.$$

Thus, by (2.3.28), (2.3.31), and the fact that A_l is self-adjoint we obtain

(2.3.32) $$\|A^{\frac{1}{2}}_l\| \leq \left(\frac{C^\gamma_l}{w_{r_l}(\pi_l)}\right)^{\frac{1}{2}} \|G\|^{\frac{1}{2}}.$$

We use (2.3.28) a second time to obtain

$$\|A^{\frac{1}{2}}_l y_l\|^2 = \langle A_l y_l, y_l\rangle \leq \frac{1}{w_{r_l}(\pi_l)} \langle (\gamma(\pi_l) * G) y_l, y_l\rangle$$

Let $E = G - \Xi$, so $\|E\| < \epsilon$. Then

(2.3.33) $$\|A^{\frac{1}{2}}_l y_l\|^2 \leq \frac{1}{w_{r_l}(\pi_l)} \left[\langle(\gamma(\pi_l) * \Xi) y_l, y_l\rangle + \langle(\gamma(\pi_l) * E) y_l, y_l\rangle\right].$$

Since Ξ is hyperextremal, (2.3.9) implies that

(2.3.34) $$\langle(\gamma(\pi_l) * \Xi) y_l, y_l\rangle = 0.$$

Furthermore, by (2.3.10), $\|y_l\| = 1$, and so we have

(2.3.35) $$\langle(\gamma(\pi_l) * E) y_l, y_l\rangle \leq \|\gamma(\pi_l) * E\|\|y_l\|$$

(2.3.36) $$\leq C^\gamma_l \|E\|$$

(2.3.37) $$\leq C^\gamma_l \epsilon.$$

Combining (2.3.33), (2.3.34), and (2.3.37), we see that

(2.3.38) $$\|A^{\frac{1}{2}}_l y_l\| \leq \left(\frac{C^\gamma_l \epsilon}{w_{r_l}(\pi_l)}\right)^{\frac{1}{2}}$$

Now (2.3.32) and (2.3.38) together yield

$$\|A_l y_l\| \leq \|A^{\frac{1}{2}}_l\|\|A^{\frac{1}{2}}_l y_l\|$$

$$\leq \frac{C^\gamma_l}{w_{r_l}(\pi_l)} \|G\|^{\frac{1}{2}} \epsilon^{\frac{1}{2}}.$$

We conclude that for each $l = 1, \ldots, 8$,

$$\|(h(\zeta_l) * \Xi + \varphi_1(\zeta_l)Y_1 + \varphi_2(\zeta_l)Y_2) y_l\| = \|(A_l + h(\zeta_l) * E) y_l\|$$

$$\leq \|A_l y_l\| + \|(h(\zeta_l) * E) y_l\|$$

$$\leq \|A_l y_l\| + C^h_l \epsilon$$

$$= \frac{C^\gamma_l}{w_{r_l}(\pi_l)} \left(\sqrt{\|\Xi\|} + \epsilon\right) \epsilon^{\frac{1}{2}} + C^h_l \epsilon$$

$$\leq E_l(\epsilon).$$

The proposition now follows by setting $\delta_l = (h(\zeta_l) * \Xi + \varphi_1(\zeta_l)Y_1 + \varphi_2(\zeta_l)Y_2) y_l$.

□

We now are able to state and prove the promised result that the non-self-adjointness of X_1 and X_2 is controlled by the distance to the dilation body. For $\epsilon > 0$, note that if we define

(2.3.39) $$\mathbf{E}(\epsilon) = (E_1(\epsilon), \ldots, E_8(\epsilon)) \in \mathbb{R}^8,$$

and if δ is as in Lemma 2.3.22, then

(2.3.40) $$\|\delta\|_2 \leq \|\mathbf{E}(\epsilon)\|,$$

where $\|.\|$ denotes the Euclidean norm on $\mathbb{R}^8$.

PROPOSITION 2.3.41. *Assume that Ξ is nonsingularly hyperextremal, and for $r = 1, 2$ let C_r be the rth 8×4 block column of $A_{\Xi,\lambda_1,\lambda_2}^{-1}$. If there exists $G \in \mathcal{D}_z$ such that $\|G - \Xi\| < \epsilon$, then the solution $\mathbf{X} = [X_1 X_2]$ of the linear system $\mathbf{X} A_{\Xi,\lambda_1,\lambda_2} = B_{\Xi,\lambda_1,\lambda_2}$ satisfies*

(2.3.42) $$\|Im X_r\|_2 \leq \|C_r\| \|\mathbf{E}(\epsilon)\|, \text{ for } r = 1, 2.$$

PROOF. Since $G \in \mathcal{D}_z$, by Proposition 2.2.40 there exist self-adjoint 4×4 matrices Y_1 and Y_2 such that

$$h(\lambda) * G + \varphi_1(\lambda) Y_1 + \varphi_2(\lambda) Y_2 \geq 0 \text{ for all } \lambda \in \partial\Omega.$$

Consequently, if we let $\mathbf{Y} = [Y_1 Y_2]$ then Lemma 2.3.22 implies that

$$\mathbf{Y} A_{\Xi,\lambda_1,\lambda_2} = B_{\Xi,\lambda_1,\lambda_2} + \delta$$

where δ is a 4×8 matrix satisfying (2.3.40). By assumption, we have

$$\mathbf{X} A_{\Xi,\lambda_1,\lambda_2} = B_{\Xi,\lambda_1,\lambda_2}.$$

Taking the difference of the above two equations, we obtain

$$\mathbf{Y} - \mathbf{X} = \delta A_{\Xi,\lambda_1,\lambda_2}^{-1}.$$

Consequently, for $r = 1, 2$

$$Y_r - X_r = \delta C_r.$$

As $Im Y_r = 0$,

$$\|Im X_r\|_2 = \|Im(Y_r - X_r)\|_2$$
$$\leq \|Y_r - X_r\|_2$$
$$\leq \|C_r\| \|\delta\|_2 \text{ by (2.3.19)}$$
$$\leq \|C_r\| \|\mathbf{E}(\epsilon)\|.$$

□

While the notion of a hyperextremal grammian is useful theoretically, it seems unlikely that a machine computation could ever generate such an object precisely. Rather, our computations in Section 2.6 lead to grammians Ξ that are "approximately" hyperextremal in the following sense.

DEFINITION 2.3.43. *Fix $\theta > 0$, $\alpha^*, \beta^* \in T_\Omega^0$ and $u^*, v^* \in \mathbb{C}^4$. A grammian $\Xi > 0$ is **approximately hyperextremal with parameter** $\Theta = (\alpha^*, \beta^*, u^*, v^*, \theta)$ provided*

(2.3.44) $$\gamma(\alpha) * \Xi \geq -\theta \text{ for all } \alpha \in T_\Omega^0$$

2.3. DILATION CONDITION FOR NONSINGULARLY HYPEREXTREMAL GRAMMIANS

and

$$\langle (\gamma(\alpha^*) * \Xi) u^*, u^* \rangle \leq \theta \tag{2.3.45}$$
$$\langle (\gamma(\beta^*) * \Xi) v^*, v^* \rangle \leq \theta. \tag{2.3.46}$$

By comparing Definition 2.3.5 with the definition just given, we see that a strongly normalized grammian Ξ is hyperextremal if and only if there exists a value of the parameter $\Theta = (\alpha^*, \beta^*, u^*, v^*, \theta)$ such that $\theta = 0$ and Ξ is approximately hyperextremal with parameter Θ. In Section 2.6, we give an example of an approximately hyperextremal grammian Ξ^* with $\theta = 10^{-17}$. We shall argue that this grammian gives rise to a counterexample to the rational dilation conjecture by exploiting an "approximate" analog of Proposition 2.3.41 (Proposition 2.3.55 below). Just as to each hyperextremal Ξ we associated the linear system $\mathbf{X} A_{\Xi,\lambda_1,\lambda_2} = B_{\Xi,\lambda_1,\lambda_2}$, we associate a linear system to each approximately hyperextremal grammian Ξ, in the following way. For $l = 1, \ldots, 8$ form the critical 4-tuples $(\pi_l, \zeta_l, y_l, r_l)$ listed Table 2.1. Define $A_{\Xi,\Theta,\lambda_1,\lambda_2}$ using equation (2.3.13) and $B_{\Xi,\Theta,\lambda_1,\lambda_2}$ using equation (2.3.14). Finally, consider the system

$$\mathbf{X} A_{\Xi,\Theta,\lambda_1,\lambda_2} = B_{\Xi,\Theta,\lambda_1,\lambda_2}. \tag{2.3.47}$$

The following definition, which parallels Definition 2.3.17, formalizes the case when (2.3.47) is nonsingular.

DEFINITION 2.3.48. *If Ξ is approximately hyperextremal with parameter Θ, we say that Ξ is **nonsingularly approximately hyperextremal with parameter Θ** if (2.3.47) is nonsingular.*

Before continuing, we record the following fact about nonsingularly approximately hyperextremal grammians.

LEMMA 2.3.49. *Assume that Ξ is nonsingularly approximately hyperextremal with parameter Θ. If $\alpha \in T_\Omega^0$ and $u \in \mathbb{C}^4$ satisfy*

$$\langle (\gamma(\alpha) * \Xi) u, u \rangle \leq \theta,$$

and $u_\lambda = [D_{\Gamma_\alpha(z)} - \Gamma_\alpha(\lambda)] u$, then

$$\langle (\gamma(\alpha(\lambda)) * \Xi) u_\lambda, u_\lambda \rangle \leq \frac{\theta}{a_{\alpha(\lambda)}} \text{ for all } \lambda \in \partial\Omega,$$

where $a_{\alpha(\lambda)}$ is the positive constant defined in (2.2.25) and (2.2.26).

PROOF. By Lemma 2.2.45 we have that

$$a_{\alpha(\lambda)} [\gamma(\alpha(\lambda)) * \Xi] = \left[\left(D_{\Gamma_\alpha(z)} - \Gamma_\alpha(\lambda) \right)^{-1} \right]^* [\gamma(\alpha)) * \Xi] \left[D_{\Gamma_\alpha(z)} - \Gamma_\alpha(\lambda) \right]^{-1}.$$

Hence,

$$\langle (\gamma(\alpha(\lambda)) * \Xi) u_\lambda, u_\lambda \rangle = \frac{1}{a_{\alpha(\lambda)}} \langle \left[(D_{\Gamma_\alpha(z)} - \Gamma_\alpha(\lambda))^{-1} \right]^* [\gamma(\alpha) * \Xi]$$
$$[D_{\Gamma_\alpha(z)} - \Gamma_\alpha(\lambda)]^{-1} u_\lambda, u_\lambda \rangle$$
$$= \frac{1}{a_{\alpha(\lambda)}} \langle \left[(D_{\Gamma_\alpha(z)} - \Gamma_\alpha(\lambda))^{-1} \right]^* [\gamma(\alpha) * \Xi] u,$$
$$[D_{\Gamma_\alpha(z)} - \Gamma_\alpha(\lambda)] u \rangle \text{ by the definition of } u_\lambda,$$
$$= \frac{1}{a_{\alpha(\lambda)}} \langle (\gamma(\alpha) * \Xi) u, u \rangle$$
$$\leq \frac{1}{a_{\alpha(\lambda)}} \theta.$$

$\square$

In order to state our distance estimates for nonsingularly approximately hyperextremal grammians, we shall require some notation. Note that the above lemma and the definition of y_l implies that for $l = 1, \ldots, 8$,

(2.3.50) $$\langle (\gamma(\pi_l) * \Xi) y_l, y_l \rangle \leq \theta_l,$$

where θ_l is given by

(2.3.51) $$\theta_l = \begin{cases} \theta \text{ for } l = 1, 2, 5, 6. \\ \frac{\theta}{a_{\alpha^*(\lambda_1)} \|u_{\lambda_1}\|^2} \text{ for } l = 3. \\ \frac{\theta}{a_{\alpha^*(\lambda_2)} \|u_{\lambda_2}\|^2} \text{ for } l = 4. \\ \frac{\theta}{a_{\beta^*(\lambda_1)} \|v_{\lambda_1}\|^2} \text{ for } l = 7. \\ \frac{\theta}{a_{\beta^*(\lambda_2)} \|v_{\lambda_2}\|^2} \text{ for } l = 8. \end{cases}$$

Also, define the error function

$$E_l^\Theta(\epsilon) = \frac{C_l^\gamma}{w_{r_l}(\pi_l)} \left(\epsilon + \frac{\theta_l}{C_l^\gamma} \right)^{\frac{1}{2}} (\|\Xi\| + \epsilon)^{\frac{1}{2}} + C_l^h \epsilon,$$

and define the error vector

(2.3.52) $$\mathbf{E}^\Theta(\epsilon) = \left(E_1^\Theta(\epsilon), \ldots, E_8^\Theta(\epsilon) \right) \in \mathbb{R}^8.$$

Finally, we require the following lemma, which plays the same role in the proof of Proposition 2.3.55 that Lemma 2.3.22 played in the proof of Proposition 2.3.41.

LEMMA 2.3.53. *Assume that Ξ is approximately nonsingularly hyperextremal with parameter Θ, and that there exists $G \in \mathcal{D}_z$ such that $\|G - \Xi\| < \epsilon$. If Y_1 and Y_2 are 4×4 matrices that satisfy*

$$h(\lambda) * G + \varphi_1(\lambda) Y_1 + \varphi_2(\lambda) Y_2 \geq 0 \text{ for all } \lambda \in \partial\Omega,$$

and $\mathbf{Y} = [Y_1 Y_2]$, then

(2.3.54) $$\mathbf{Y} A_{\Xi,\Theta,\lambda_1,\lambda_2} = B_{\Xi,\Theta,\lambda_1,\lambda_2} + \delta,$$

where δ is a 4×8 matrix whose l^{th} column, δ_l, satisfies $\|\delta_l\|_2 \leq E_l^\Theta(\epsilon)$.

2.3. DILATION CONDITION FOR NONSINGULARLY HYPEREXTREMAL GRAMMIANS

PROOF. The proof proceeds exactly as the proof of Lemma 2.3.22, with the exception that (2.3.34) is replaced by

$$\langle (\gamma(\pi_l) * \Xi) y_l, y_l \rangle \leq \theta_l,$$

according to (2.3.50). □

The proposition now follows in exactly the same manner as Proposition 2.3.41.

PROPOSITION 2.3.55. *Assume that Ξ is approximately nonsingularly hyperextremal with parameter Θ, and for $r = 1, 2$ let C_r be the rth 8×4 block column of $A_{\Xi,\Theta,\lambda_1,\lambda_2}^{-1}$. If there exists $G \in \mathcal{D}_z$ such that $\|G - \Xi\| < \epsilon$, then the solution $\mathbf{X} = [X_1 X_2]$ of the linear system $\mathbf{X} A_{\Xi,\Theta,\lambda_1,\lambda_2} = B_{\Xi,\Theta,\lambda_1,\lambda_2}$ satisfies*

$$\text{(2.3.56)} \qquad \|\text{Im} X_r\|_2 \leq \|C_r\| \|\mathbf{E}^\Theta(\epsilon)\|, \text{ for } r = 1, 2.$$

We close this section with a second distance estimate, based on the fact that for the grammians Ξ that we consider in Section 2.6, the solution of the linear system (2.3.47) yields X_1 and X_2 which are not self-adjoint on the two dimensional subspace spanned by u^* and v^*. Accordingly, for $X \in \mathbb{C}^{4,4}$ define a sesquilinear form, κ_X on $\mathbb{C}^4$ by setting

$$\kappa_X(a, b) = \langle Xa, b \rangle - \langle a, Xb \rangle.$$

To state our result about self-adjointness on pairs of vectors, we use the following notation. For $\alpha \in T_\Omega^0$ define $M_\alpha \in \mathbb{C}^{2,2}$ by

$$\text{(2.3.57)} \qquad M_\alpha = \begin{pmatrix} \varphi_1(\alpha_1) & \varphi_2(\alpha_1) \\ \varphi_1(\alpha_2) & \varphi_2(\alpha_2) \end{pmatrix}^{-1}.$$

For $r = 1, 2$ and $\alpha \in T_\Omega$, let

$$\text{(2.3.58)} \qquad \nu_r(\alpha) = r^{th} \text{ row of } M_\alpha.$$

PROPOSITION 2.3.59. *Assume that Ξ is approximately nonsingularly hyperextremal with parameter Θ. If there exists $G \in \mathcal{D}_z$ such that $\|G - \Xi\| < \epsilon$, then the solution $\mathbf{X} = [X_1 X_2]$ of the linear system $\mathbf{X} A_{\Xi,\Theta,\lambda_1,\lambda_2} = B_{\Xi,\Theta,\lambda_1,\lambda_2}$ satisfies*

$$\text{(2.3.60)} \quad |\kappa_{X_r}(u^*, v^*)| \leq \|\nu_r(\alpha^*)\| \left(E_1^\Theta(\epsilon)^2 + E_2^\Theta(\epsilon)^2 \right)^{\frac{1}{2}}$$

$$+ \|\nu_r(\beta^*)\| \left(E_5^\Theta(\epsilon)^2 + E_6^\Theta(\epsilon)^2 \right)^{\frac{1}{2}} \text{ for } r = 1, 2.$$

PROOF. Since $G \in \mathcal{D}_z$, by (2.2.41) there exist self-adjoint matrices Y_1 and Y_2 such that

$$h(\lambda) * G + \varphi_1(\lambda) Y_1 + \varphi_2(\lambda) Y_2 \geq 0 \text{ for all } \lambda \in \partial \Omega.$$

Consequently, if we let $\mathbf{Y} = [Y_1 Y_2]$ then we see from Lemma 2.3.53 that

$$\mathbf{Y} A_{\Xi,\Theta,\lambda_1,\lambda_2} = B_{\Xi,\Theta,\lambda_1,\lambda_2} + \delta,$$

where δ is a 4×8 matrix whose columns satisfy $\|\delta_l\| \leq E_l^\Theta(\epsilon)$. Let $\mathbf{Z} = \mathbf{Y} - \mathbf{X}$, so that $\mathbf{Z}$ satisfies

$$\text{(2.3.61)} \qquad \mathbf{Z} A_{\Xi,\Theta,\lambda_1,\lambda_2} = \delta.$$

Now recall from (2.3.13) that the first two columns of (2.3.61) are

$$[Z_1 Z_2] \begin{pmatrix} \varphi_1(\zeta_1) y_1 & \varphi_1(\zeta_2) y_2 \\ \varphi_2(\zeta_1) y_1 & \varphi_2(\zeta_2) y_2 \end{pmatrix} = [\delta_1 \delta_2].$$

Recalling that $y_1 = y_2 = u^*$ and $\zeta_1 = \alpha_1^*$ and $\zeta_2 = \alpha_2^*$, we obtain

$$[Z_1 Z_2] \begin{pmatrix} \varphi_1(\alpha_1^*)u^* & \varphi_1(\alpha_2^*)u^* \\ \varphi_2(\alpha_1^*)u^* & \varphi_2(\alpha_2^*)u^* \end{pmatrix} = [\delta_1 \delta_2].$$

which can be expressed as

$$\begin{pmatrix} \varphi_1(\alpha_1^*) & \varphi_2(\alpha_1^*) \\ \varphi_1(\alpha_2^*) & \varphi_2(\alpha_2^*) \end{pmatrix} \begin{pmatrix} Z_1 u^* \\ Z_2 u^* \end{pmatrix} = \begin{pmatrix} \delta_1 \\ \delta_2 \end{pmatrix}.$$

Hence, viewing the matrix M_α of (2.3.57) as a block operator on $\mathbb{C}^4 \oplus \mathbb{C}^4$ we see that

$$\begin{pmatrix} Z_1 u^* \\ Z_2 u^* \end{pmatrix} = M_{\alpha^*} \begin{pmatrix} \delta_1 \\ \delta_2 \end{pmatrix}.$$

Likewise, we obtain that

$$\begin{pmatrix} Z_1 v^* \\ Z_2 v^* \end{pmatrix} = M_{\beta^*} \begin{pmatrix} \delta_5 \\ \delta_6 \end{pmatrix}.$$

Since Y_1 and Y_2 are self-adjoint, $\kappa_{Y_r} = 0$ for $r = 1, 2$ and thus

$$\kappa_{X_r}(u^*, v^*) = \kappa_{Y_r - Z_r}(u^*, v^*)$$
$$= -\kappa_{Z_r}(u^*, v^*)$$
$$= \langle u^*, Z_r v^* \rangle - \langle Z_r u^*, v^* \rangle$$
$$= \left\langle u^*, \nu_r(\beta^*) \begin{pmatrix} \delta_5 \\ \delta_6 \end{pmatrix} \right\rangle - \left\langle \nu_r(\alpha^*) \begin{pmatrix} \delta_1 \\ \delta_2 \end{pmatrix}, v^* \right\rangle.$$

Recalling that $\|u^*\| = \|v^*\| = 1$ and $\|\delta_l\| \leq E_l^\Theta(\epsilon)$ we obtain that for $r = 1, 2$,

$$|\kappa_{X_r}(u^*, v^*)| \leq \left|\left\langle \nu_r(\alpha^*) \begin{pmatrix} \delta_1 \\ \delta_2 \end{pmatrix}, v^* \right\rangle\right| + \left|\left\langle u^*, \nu_r(\beta^*) \begin{pmatrix} \delta_5 \\ \delta_6 \end{pmatrix} \right\rangle\right|$$

$$\leq \|\nu_r(\alpha^*)\| \left(E_1^\Theta(\epsilon)^2 + E_2^\Theta(\epsilon)^2\right)^{\frac{1}{2}} + \|\nu_r(\beta^*)\| \left(E_5^\Theta(\epsilon)^2 + E_6^\Theta(\epsilon)^2\right)^{\frac{1}{2}}.$$

$\square$

2.4. Analysis of Dilation Extremal Grammians

In the previous section, we defined grammians that are hyperextremal with respect to condition (2.2.39). We used the hyperextremality to obtain specific estimates of the distance between these grammians and the dilation body, $\mathcal{D}_z$. In this section, we describe an alternate approach, a method that is based on analyzing grammians that are extremal with respect to the dilation condition (2.2.41). We define *dilation extremal* grammians (Definition 2.4.3) and show how to attach a system of linear inequalities to such grammians. The feasibility of the system of linear inequalities gives rise to an estimate of the distance between a dilation extremal grammian and the boundary of the dilation body. Lemma 2.4.21 gives such an estimate in radial distance (i.e. the distance to $\partial \mathcal{D}_z$ along a ray emanating from the identity matrix), while Proposition 2.4.29 gives an estimate in the operator norm.

Before continuing, we indicate how the results of this section will be applied in Section 2.6 to justify a counterexample to the rational dilation conjecture. We assume a strongly normalized grammian Ξ is given and consider the optimization problem

(2.4.1) $$\tau^* = \max\{\tau : 1 - \tau(1 - \Xi) \in \mathcal{D}_z\}.$$

2.4. ANALYSIS OF DILATION EXTREMAL GRAMMIANS

Now, if $\Xi \in \partial \mathcal{S}_z$ (e.g. if Ξ is hyperextremal) and $\mathcal{D}_z = \mathcal{S}_z$ (i.e. the rational dilation conjecture is true), then necessarily $1 - \tau^* = 0$. In Section 2.6 we present a specific hyperextremal grammian Ξ. For this hyperextremal grammian, it turns out that $1 - \tau^*$ is strictly positive. Accordingly, in Section 2.6 we will combine the fact that $1 - \tau^*$ is strictly positive with the distance estimates derived in this section to obtain a lower bound between Ξ and the dilation body. These estimates result in a second verification that the grammian Ξ provides a counterexample to the rational dilation conjecture, a verification that is independent of the methods employed in the previous section; in particular, the verification is independent of the hyperextremality of Ξ.

The distance estimates are based on the striking empirical observation that solving (2.4.1) gives rise to grammians

$$(2.4.2) \qquad \Pi = 1 - \tau^*(1 - \Xi)$$

that are dilation extremal in the sense of the following definition.

DEFINITION 2.4.3. *If $\Pi > 0$ is a strongly normalized grammian, we say that Π is **dilation extremal** provided there exist self-adjoint matrices $X_1, \ldots, X_m$ such that*

$$(2.4.4) \qquad h(\lambda) * \Pi + \sum_{r=1}^{m} \varphi_r(\lambda) X_r \geq 0 \text{ for all } \lambda \in \partial \Omega,$$

and

$$(2.4.5) \qquad \ker \left[h(\lambda) * \Pi + \sum_{r=1}^{m} \varphi_r(\lambda) X_r \right] \neq \{0\} \text{ for all } \lambda \in \partial \Omega.$$

Thus, if $\Pi > 0$ is a strongly normalized grammian, then Π is dilation extremal provided $\Pi \in \mathcal{D}_z$ (this is guaranteed by (2.4.4) and Proposition 2.2.40), and, in addition, (2.4.4) holds sharply at each $\lambda \in \partial \Omega$. Note that the notion of a dilation extremal grammian, which is based on Proposition 2.2.40, is qualitatively stronger than the notion of extremal grammian, which is based on Proposition 2.2.38 and was formalized in Definition 2.3.2. In particular, for a dilation extremal, the condition (2.2.41) must be met sharply for all $\lambda \in \partial \Omega$, while for an extremal grammian, condition (2.2.39) is required to be met sharply only for a single $\alpha \in T_\Omega^0$.

However, Proposition 2.2.46 reveals that if condition (2.2.39) is met sharply for some $\alpha \in T_\Omega$, then (2.2.39) is met sharply for $\alpha(\lambda) \in T_\Omega$, for all $\lambda \in \partial \Omega$. This observation allows us to prove the following connection between the notions of dilation extremality and extremality.

PROPOSITION 2.4.6. *If $G > 0$ is strongly normalized, G is extremal, and $G \in \mathcal{D}_z$ then G is dilation extremal.*

PROOF. Since $G \in \mathcal{D}_z$, by (2.2.41) there exists self-adjoint matrices $X_1, \ldots, X_m$ such that

$$(2.4.7) \qquad h(\lambda) * G + \sum_{s=1}^{m} \varphi_s(\lambda) X_s \geq 0 \text{ for all } \lambda \in \partial \Omega.$$

Since G is extremal, there exist $\alpha^* \in T_\Omega^0$ and $u^* \in \mathbb{C}^n$ such that $[\gamma(\alpha^*) * G]\, u^* = 0$. For $\lambda \in \partial\Omega$, let $u_\lambda = \left[D_{\Gamma_{\alpha^*}(z)} - \Gamma_{\alpha^*}(\lambda)\right] u^*$. By Proposition 2.2.43 and Proposition 2.2.46, we have

$$0 = [\gamma(\alpha(\lambda)) * G]\, u_\lambda = \sum_{r=0}^{m} w_r(\alpha(\lambda)) \left[h(\alpha(\lambda)_r) * G + \sum_{s=1}^{m} \varphi_s(\alpha(\lambda)_r) X_s \right] u_\lambda.$$

Now (2.4.7) implies that

$$\left[h(\alpha(\lambda)_r) * G + \sum_{s=1}^{m} \varphi_s(\alpha(\lambda)_r) X_s \right] \geq 0, \text{ for } r = 0, \ldots, m,$$

and since $w_r(\alpha) > 0$ for all $\alpha \in T_\Omega$ we obtain

$$\left[h(\alpha(\lambda)_r) * G + \sum_{s=1}^{m} \varphi_s(\alpha(\lambda)_r) X_s \right] u_\lambda = 0, \text{ for } r = 0, \ldots, m.$$

In particular, if $\lambda \in \partial_r$ then $\alpha(\lambda)_r = \lambda$, and thus

$$\left[h(\lambda) * G + \sum_{s=1}^{m} \varphi_s(\lambda) X_s \right] u_\lambda = 0.$$

We conclude that G is dilation extremal. $\square$

Just as in Section 2.3, where we attached the system of linear *equations* (2.3.16) to each hyperextremal grammian, we associate a system of linear *inequalities* to each dilation extremal grammian Π in the following way. For $\lambda \in \partial\Omega$ and $x \in \mathbb{C}^n$ define a real linear functional $L_{\lambda,x}$ on the real vector space of m-tuples $\mathbf{Z} = (Z_1, \ldots, Z_m)$ of $n \times n$ self-adjoint matrices by

$$(2.4.8) \qquad L_{\lambda,x}(\mathbf{Z}) = \left\langle \left(\sum_{r=1}^{m} \varphi_r(\lambda) Z_r \right) x, x \right\rangle.$$

Let $\eta \in \mathbb{R}$, and for each $\lambda \in \partial\Omega$ choose $x_\lambda \in \mathbb{C}^n$ such that

$$(2.4.9) \qquad x_\lambda \in \ker\left[h(\lambda) * \Pi + \sum_{r=1}^{m} \varphi_r(\lambda) X_r \right] \text{ and } \|x_\lambda\| = 1.$$

In terms of these choices, we consider the system of linear inequalities,

$$(2.4.10) \qquad L_{\lambda,x_\lambda}(\mathbf{Z}) \geq \eta, \text{ for all } \lambda \in \partial\Omega.$$

Our analysis of dilation extremals will begin with the following lemma whose proof will use the following constants:

$$(2.4.11) \qquad C^p = \min_{\lambda \in \partial\Omega, i=1,\ldots,n} P_\lambda(z_i),$$

$$(2.4.12) \qquad C^h = \max_{\lambda \in \partial\Omega, i=1,\ldots,n} |H_\lambda(z_i)|.$$

LEMMA 2.4.13. $I_n \in int(\mathcal{D}_z)$.

PROOF. For $r = 1, \ldots m$, define the $n \times n$ diagonal matrix

$$(2.4.14) \qquad D_r = D_{\varphi_r(z)},$$

2.4. ANALYSIS OF DILATION EXTREMAL GRAMMIANS

a matrix whose ii^{th} entry is $\varphi_r(z_i)$. By the definition of $h(\lambda)$, $h(\lambda) * I = D_{\text{Re}H_\lambda(z)}$. Hence, by (2.2.4), we have

$$h(\lambda) * I + \sum_{r=1}^{m} \varphi_r(\lambda) D_r = D_{\text{Re}H_\lambda(z)} + \sum_{r=1}^{m} \varphi_r(\lambda) D_r$$
$$= D_{P_\lambda(z)}.$$

For all $\lambda \in \partial\Omega$, $P_\lambda(z) > 0$. Thus, we obtain

$$(2.4.15) \quad h(\lambda) * I + \sum_{r=1}^{m} \varphi_r(\lambda) D_r \geq \min_{\lambda \in \partial\Omega, i=1,\ldots,n} P_\lambda(z_i) = C^p \text{ for all } \lambda \in \partial\Omega.$$

Finally, if E is an $n \times n$ self-adjoint matrix and $\|E\| \leq \frac{C^p}{C^h}$ then

$$(2.4.16) \quad h(\lambda) * (I + E) + \sum_{r=1}^{m} \varphi_r(\lambda) D_r \geq C^p - C^h \|E\| \geq 0.$$

Thus $I + E \in \mathcal{D}_z$, and the proposition follows. □

Next, using Lemma 2.4.13 and the convexity of $\mathcal{D}_z$, we obtain an estimate of the positivity of the dilation condition (2.2.41) for grammians G which are elements of $int(\mathcal{D}_z)$. For G a strongly normalized grammian define the ray of grammians

$$(2.4.17) \quad G_\tau = (1-\tau)I + \tau G, \tau \geq 0.$$

LEMMA 2.4.18. *If $G \in \mathcal{D}_z$, then there exist self-adjoint $n \times n$ matrices $X_1, \ldots, X_m$ such that*

$$h(\lambda) * G_t + \sum_{r=1}^{m} \varphi_r(\lambda) X_r \geq (1-t) C^p \text{ for } 0 \leq t \leq 1.$$

PROOF. If $D_1, \ldots, D_m$ are as in (2.4.14) then Lemma 2.4.13 implies

$$(2.4.19) \quad h(\lambda) * I + \sum_{r=1}^{m} \varphi_r(\lambda) D_r \geq C^p.$$

Since $G \in \mathcal{D}_z$ by Proposition 2.2.40 there exist self-adjoint matrices $Y_1 \ldots Y_m$ such that

$$(2.4.20) \quad h(\lambda) * G + \sum_{r=1}^{m} \varphi_r(\lambda) Y_r \geq 0.$$

Hence, taking a convex combination of (2.4.19) and (2.4.20), we obtain

$$h(\lambda) * G_t + \sum_{r=1}^{m} \varphi_r(\lambda) \left((1-t) D_r + t Y_r\right) \geq (1-t) C^p.$$

□

LEMMA 2.4.21. *Assume that Π is dilation extremal and that for each $\lambda \in \partial\Omega$ $x_\lambda \in \mathbb{C}^n$ satisfies (2.4.9). If there exists $\tau > 1$ such that*

$$G = (1-\tau)I + \tau\Pi \in \mathcal{D}_z,$$

then the system of linear inequalities, (2.4.10), is solvable with $\eta = \frac{\tau-1}{\tau} C^p$.

PROOF. Let $t = \frac{1}{\tau}$. Then
$$\Pi = (1-t)I + tG.$$
By Lemma 2.4.18, there exist self-adjoint matrices $Y_1, \ldots Y_m$ such that

(2.4.22) $$h(\lambda) * \Pi + \sum_{r=1}^{m} \varphi_r(\lambda) Y_r \geq (1-t)C^p.$$

Since $\Pi \in \mathcal{D}_z$, by Proposition 2.2.40 there exist self-adjoint matrices $X_1, \ldots X_m$ such that

(2.4.23) $$h(\lambda) * \Pi + \sum_{r=1}^{m} \varphi_r(\lambda) X_r \geq 0 \text{ for all } \lambda \in \partial \Omega.$$

Taking a convex combination of (2.4.22) and (2.4.23) with $0 < s < 1$, we obtain

(2.4.24) $$h(\lambda) * \Pi + s \sum_{r=1}^{m} \varphi_r(\lambda) Y_r + (1-s) \sum_{r=1}^{m} \varphi_r(\lambda) X_r \geq s(1-t)C^p$$

and hence

(2.4.25) $$h(\lambda) * \Pi + \sum_{r=1}^{m} \varphi_r(\lambda) X_r + s \sum_{r=1}^{m} \varphi_r(\lambda) (Y_r - X_r) \geq s(1-t)C^p$$

We multiply by x_λ and take the inner product with x_λ to obtain

$$\left\langle \left[h(\lambda) * \Pi + \sum_{r=1}^{m} \varphi_r(\lambda) X_r \right] x_\lambda, x_\lambda \right\rangle + s \left\langle \sum_{r=1}^{m} \varphi_r(\lambda) (Y_r - X_r) x_\lambda, x_\lambda \right\rangle \geq s(1-t)C^p.$$

Since Π is a dilation extremal, the first inner product is zero, and so we have

$$\left\langle \sum_{r=1}^{m} \varphi_r(\lambda) (Y_r - X_r) x_\lambda, x_\lambda \right\rangle \geq (1-t)C^p.$$

The proposition follows by noting that $1 - t = \frac{\tau - 1}{\tau}$ and setting $Z_r = Y_r - X_r$. □

Lemmas 2.4.18 and 2.4.21 followed from the consideration of the rays of grammians defined in (2.4.17) for $\tau > 0$. To obtain estimates of the distance to the dilation body, $\mathcal{D}_z$, consideration of rays (i.e. "radial" distance) is not sufficient. Lemma 2.4.27 below provides a connection between rays of grammians and balls of grammians. To state this lemma, we first make two definitions. If $G > 0$ is strongly normalized and $\epsilon > 0$, let $B_\epsilon(G)$ be the open ball

$$B_\epsilon(G) = \{H > 0 : H \text{ strongly normalized}, \|G - H\| < \epsilon\}.$$

Also, we define the constant

(2.4.26) $$C^{\text{radius}} = \sup\{r : B_r(I) \subseteq \mathcal{D}_z\}.$$

Note that by (2.4.16), $C^{\text{radius}} \geq \frac{C^p}{C^h}$.

LEMMA 2.4.27. *If $\Xi > 0$ is strongly normalized, $G \in \mathcal{D}_z$, and for $\tau \geq 0$, Ξ_τ denotes the ray of grammians*

(2.4.28) $$\Xi_\tau = (1-\tau)I + \tau \Xi,$$

then $B_{(1-\tau)C^{\text{radius}} - \tau \|G - \Xi\|}(\Xi_\tau) \subseteq \mathcal{D}_z$.

2.4. ANALYSIS OF DILATION EXTREMAL GRAMMIANS

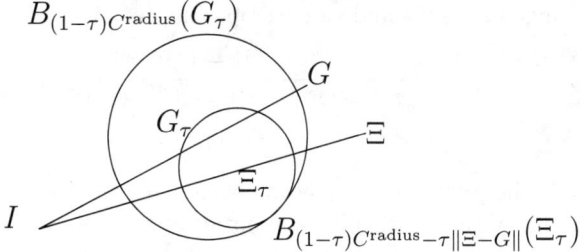

FIGURE 2.1. Illustration of Lemma 2.4.27

PROOF. By the definition of C^{radius},
$$B_{C^{\mathrm{radius}}}(I) \subseteq \mathcal{D}_z.$$
Since $\mathcal{D}_z$ is convex, and by assumption $G \in \mathcal{D}_z$, we obtain
$$B_{(1-\tau)C^{\mathrm{radius}}}(G_\tau) = (1-\tau) B_{C^{\mathrm{radius}}}(I) + \tau G \subseteq \mathcal{D}_z,$$
where G_τ is defined in (2.4.17). Noting that $\|\Xi_\tau - G_\tau\| = \tau \|\Xi - G\|$ and using the triangle inequality, we conclude
$$B_{(1-\tau)C^{\mathrm{radius}} - \tau\|\Xi - G\|}(\Xi_\tau) = B_{(1-\tau)C^{\mathrm{radius}} - \|\Xi_\tau - G_\tau\|}(\Xi_\tau) \subseteq \mathcal{D}_z.$$
See Figure 2.1. □

Using Lemmas 2.4.21 and 2.4.27, we obtain the following estimate of the distance between a dilation extremal grammian and the boundary of the dilation body.

PROPOSITION 2.4.29. *Let $\epsilon, \tau^* > 0$ with $\tau^* < 1$ and $\epsilon < \frac{1-\tau^*}{\tau^*} C^{\mathrm{radius}}$. Assume that Ξ is a strongly normalized grammian, that Ξ_{τ^*} is dilation extremal, and that for each $\lambda \in \partial \Omega$, x_λ satisfies (2.4.9). If there exists $G \in \mathcal{D}_z$ such that $\|G - \Xi\| < \epsilon$, then (2.4.10) is solvable with $\eta = \frac{\delta}{\rho + \delta} C^p$ where*

(2.4.30) $$\delta = (1-\tau^*) C^{\mathrm{radius}} - \tau^* \epsilon,$$

and

(2.4.31) $$\rho = \|\Xi - I\| \tau^*.$$

PROOF. Noting that $\|G - \Xi\| < \epsilon$, we see from Lemma 2.4.27 that
$$B_{(1-\tau^*)C^{\mathrm{radius}} - \tau^* \epsilon}(\Xi_{\tau^*}) \subseteq \mathcal{D}_z.$$
Consequently,

(2.4.32) $$\left[-\left(\frac{(1-\tau^*)C^{\mathrm{radius}} - \tau^* \epsilon}{\tau^* \|\Xi - I\|} \right) \right] I + \left(1 + \frac{(1-\tau^*)C^{\mathrm{radius}} - \tau^* \epsilon}{\tau^* \|\Xi - I\|} \right) \Xi_{\tau^*}$$
$$= \Xi_{\tau^*} + \frac{(1-\tau^*)C^{\mathrm{radius}} - \tau^* \epsilon}{\|\Xi - I\|} (\Xi - I) \in \mathcal{D}_z.$$

We now apply Lemma 2.4.21 with $\Pi = \Xi_{\tau^*}$ and
$$\tau = 1 + \frac{(1-\tau^*)C^{\mathrm{radius}} - \tau^* \epsilon}{\tau^* \|\Xi - I\|} > 1$$
to conclude that
$$L_{\lambda, x_\lambda}(\mathbf{Z}) \geq \frac{\tau - 1}{\tau} C^p \text{ for all } \lambda \in \partial \Omega.$$

Simplifying the right-hand side we obtain

$$\frac{\tau - 1}{\tau} = \frac{(1 - \tau^*)C^{\text{radius}} - \tau^*\epsilon}{\tau^*\|\Xi - I\| + (1 - \tau^*)C^{\text{radius}} - \tau^*\epsilon}$$

□

While the properties of dilation extremal grammians are useful theoretically, it seems unlikely that a machine computation could ever generate a dilation extremal grammian precisely. Rather, using algorithms outlined in the next section, the authors obtained a number of grammians Π (and associated self-adjoint matrices $X_1, \ldots, X_m$) that are "approximately" dilation extremal in the following sense.

DEFINITION 2.4.33. $\Pi > 0$ is **approximately dilation extremal with parameter** $\xi = (\xi_1, \xi_2)$ with $\xi_i > 0$, provided there exists self-adjoint matrices $X_1, \ldots, X_m$ such that:

(2.4.34) $$h(\lambda) * \Pi + \sum_{r=1}^{m} \varphi_r(\lambda) X_r \geq -\xi_1 \text{ for all } \lambda \in \partial\Omega,$$

and

(2.4.35) $$h(\lambda) * \Pi + \sum_{r=1}^{m} \varphi_r(\lambda) X_r \leq \xi_2 \text{ for all } \lambda \in \partial\Omega.$$

For these grammians satisfying the conditions of Definition 2.4.33, we formulate the following analogue of Proposition 2.4.29.

PROPOSITION 2.4.36. *Let $\epsilon, \tau^* > 0$ with $\tau^* < 1$ and $\epsilon < \frac{1-\tau^*}{\tau^*} C^{\text{radius}}$. Assume that $\Xi > 0$ is a strongly normalized grammian and that Ξ_{τ^*} is "approximately" dilation extremal with parameter ξ. For each $\lambda \in \partial\Omega$, let x_λ be a unit eigenvector corresponding to the minimum eigenvalue of $h(\lambda) * \Xi_{\tau^*} + \sum_{r=0}^{m} \varphi_r(\lambda) X_r$. If there exists $G \in \mathcal{D}_z$ such that $\|\Xi - G\| < \epsilon$, then the system of inequalities (2.4.10) is solvable with $\eta = \frac{\delta}{\rho + \delta} C^p - \xi_2$, where*

(2.4.37) $$\delta = (1 - \tau^*) C^{\text{radius}} - \tau^*\epsilon,$$

and

(2.4.38) $$\rho = \|\Xi - I\|\tau^*.$$

PROOF. Noting that $\|G - \Xi\| < \epsilon$, we see from Lemma 2.4.27 that

$$B_{(1-\tau^*)C^{\text{radius}} - \tau^*\epsilon}(\Xi_{\tau^*}) \subseteq \mathcal{D}_z.$$

Consequently,

$$\Xi_{\tau^*} + \frac{(1 - \tau^*)C^{\text{radius}} - \tau^*\epsilon}{\|\Xi - I\|}(\Xi - I) \in \mathcal{D}_z.$$

Define

$$\sigma = \frac{(1 - \tau^*)C^{\text{radius}} - \tau^*\epsilon}{\|\Xi - I\|} = \frac{\delta \tau^*}{\rho}.$$

Then

$$\Xi_{\tau^*} = \left(1 - \frac{\tau^*}{\tau^* + \sigma}\right)I + \left(\frac{\tau^*}{\tau^* + \sigma}\right)\Xi_{\tau^* + \sigma}.$$

Hence, by Lemma 2.4.18, there exist self-adjoint matrices $Y_1, \ldots Y_m$ such that

$$(2.4.39) \qquad h(\lambda) * \Xi_{\tau^*} + \sum_{r=1}^{m} \varphi_r(\lambda) Y_r \geq \left(1 - \frac{\tau^*}{\tau^* + \sigma}\right) C^p.$$

Thus,

$$h(\lambda) * \Xi_{\tau^*} + \sum_{r=1}^{m} \varphi_r(\lambda) X_r + \sum_{r=1}^{m} \varphi_r(\lambda)(Y_r - X_r) \geq \left(1 - \frac{\tau^*}{\tau^* + \sigma}\right) C^p.$$

Multiplying by x_λ and taking inner products, we obtain

$$(2.4.40)$$
$$\left\langle \left[h(\lambda) * \Xi_{\tau^*} + \sum_{r=1}^{m} \varphi_r(\lambda) X_r\right] x_\lambda, x_\lambda \right\rangle + \left\langle \left[\sum_{r=1}^{m} \varphi_r(\lambda)(Y_r - X_r)\right] x_\lambda, x_\lambda \right\rangle$$
$$\geq \left(1 - \frac{\tau^*}{\tau^* + \sigma}\right) C^p.$$

By assumption, for $\lambda \in \partial \Omega$,

$$\left\langle \left[h(\lambda) * \Xi + \sum_{r=1}^{m} \varphi_r(\lambda) X_r\right] x_\lambda, x_\lambda \right\rangle \leq \xi_2.$$

Thus,

$$\left\langle \left[\sum_{r=1}^{m} \varphi_r(\lambda)(Y_r - X_r)\right] x_\lambda, x_\lambda \right\rangle \geq \left(\frac{\sigma}{\sigma + \tau^*}\right) C^p - \xi_2 \text{ for } \lambda \in \partial\Omega.$$

The proposition follows by setting $Z_r = Y_r - X_r$ and noting that $\frac{\sigma}{\sigma+\tau^*} = \frac{\delta}{\rho+\delta}$. □

We remark that in the above proposition the lower bound η for the system (2.4.10) depends only on ξ_2 and not ξ_1.

2.5. Algorithms

In this section, we describe algorithms for for constructing and analyzing the theoretical objects studied in the previous two sections, namely: extremal, hyperextremal, and dilation extremal grammians. In Section 2.5.1, we present an algorithm for constructing extremal grammians. In Section 2.5.2, we describe a method that in the case $m = 2$ and $n = 4$, when seeded with an extremal grammian, yields a nonsingularly approximately hyperextremal grammian. In Section 2.5.3, we provide a semidefinite programming formulation of the dilation step (2.4.1) that leads to dilation extremal grammians. Implementation of these algorithms requires a method for computing the functions φ_r, w, H_λ, and Γ_α with high precision. In Section 2.5.4, we describe such a method. Finally, in Section 2.5.5, we summarize the procedures that we used to solve the various optimization problems over T_Ω^0 and $\partial \Omega$ that arise in the computations.

2.5.1. Constructing Extremal Elements of $\mathcal{S}_z$.

Our procedure for computing extremal elements of $\mathcal{S}_z$ exploits the convexity and compactness of $\mathcal{S}_z$ (cf. Proposition 2.2.42) to define a "step" along a ray in $\mathcal{S}_z$ in the following way. Let I_n denote the $n \times n$ identity matrix. Choose a non-zero self-adjoint direction matrix Δ satisfying $\Delta_{ii} = 0$, so that for $t \in \mathbb{R}$, $I_n - t\Delta$ is strongly normalized. We define the step, t^*, in the direction Δ, by

$$(2.5.1) \qquad t^* = \max\{t : I_n - t\Delta \in \mathcal{S}_z\}.$$

Recalling from Lemma 2.4.13 that $I_n \in int(\mathcal{D}_z)$ and that $\mathcal{D}_z \subseteq \mathcal{S}_z$, we see immediately that $I_n \in int(\mathcal{S}_z)$. Since $\mathcal{S}_z$ is compact (Proposition 2.2.42), we conclude that t^* is well-defined, $t^* > 0$, and $I_n - t^*\Delta \in \partial \mathcal{S}_z$. Furthermore, since $\mathcal{S}_z$ is convex (Proposition 2.2.42), any element of $\partial \mathcal{S}_z$ has the form $I_n - t^*\Delta$ for some direction matrix Δ and with t^* defined in (2.5.1). Finally, in view of Proposition 2.2.49 there must exist an $\alpha^* \in T_\Omega^0$ and a unit vector $u^* \in \mathbb{C}^n$ such that $[\gamma(\alpha^*) * (I_n - t^*\Delta)]\, u^* = 0$.

We now show how to reformulate the definition of the step t^* into an optimization problem over T_Ω^0. Note that (2.2.50) implies that

$$t^* = \max\{t : D_{Re\Gamma_\alpha(z)} - t\,(\gamma(\alpha) * \Delta) \geq 0, \text{ for all } \alpha \in T_\Omega^0\}.$$

Conjugating by $D_{Re\Gamma_\alpha(z)}^{-\frac{1}{2}}$ and defining an $n \times n$ self-adjoint matrix-valued function W on T_Ω^0 by

$$W(\alpha) = D_{Re\Gamma_\alpha(z)}^{-\frac{1}{2}}(\gamma(\alpha) * \Delta) D_{Re\Gamma_\alpha(z)}^{-\frac{1}{2}},$$

we obtain

$$t^* = \max\{t : tW(\alpha) \leq 1 \text{ for all } \alpha \in T_\Omega^0\}.$$

For P a positive definite $n \times n$ matrix, let $\lambda_{\max}(P)$ denote the largest eigenvalue of P. Since $W(\alpha)$ is self-adjoint, $tW(\alpha) \leq 1$ if and only if $\lambda_{\max}(W(\alpha)) \leq \frac{1}{t}$. Hence, we obtain the step t^* by solving the following optimization problem:

$$(2.5.2) \qquad \frac{1}{t^*} = \max_{\alpha \in T_\Omega^0} \lambda_{\max}(W(\alpha)).$$

Note that the objective function $\lambda_{\max}(W(\alpha))$ in this optimization problem is a function of the m variables defining a point $\alpha \in T_\Omega^0$. If instead of condition (2.2.50), we had used condition (2.2.39) to define the step, the objective function would have been a function of the $m+1$ variables defining a point in T_Ω. When m is small, the reduction from $m+1$ to m variables (i.e. from T_Ω to T_Ω^0) is a significant improvement. The algorithm we employ to solve (2.5.2) is described in Section 2.5.5.

Now, the grammian $I_n - t^*\Delta$ is extremal in the sense of Definition 2.3.2. In the next subsection, we describe a procedure designed to find hyperextremal grammians via a second step initiated from $I_n - t^*\Delta$. To compute the second step, we shall require that $I_n - t^*\Delta$ satisfies certain "generic" criteria formalized in the following definition.

DEFINITION 2.5.3. *If $G > 0$ is strongly normalized, we say that G is **generically extremal** if*

(1) *Conditions (2.3.3) and (2.3.4) hold, i.e.*

$$\gamma(\alpha) * G \geq 0 \text{ for all } \alpha \in T_\Omega^0$$

and

> There exists $\alpha^* \in T_\Omega^0$ and $u^* \in \mathbb{C}^n$ with $\|u^*\| = 1$ such that
> $$(\gamma(\alpha^*) * G) u^* = 0.$$

(2) If $\alpha \in T_\Omega^0$ and $\alpha \neq \alpha^*$, then $\gamma(\alpha) * G > 0$.
(3) $dim\left[ker\left(\gamma(\alpha^*) * G\right)\right] = 1$.
(4) The Hessian of the function $\lambda_{\min}\left(\gamma(\alpha) * G\right)$ is strictly positive definite at α^*, i.e. $\left[\frac{\partial^2}{\partial \alpha_r \partial \alpha_s} \lambda_{\min}\left(\gamma(\alpha) * G\right)\right]\Big|_{\alpha=\alpha^*} > 0$.

We remark that since $\gamma(\alpha) * G$ is a real-analytic matrix-valued function, condition (3) in the definition implies that $\lambda_{\min}\left(\gamma(\alpha) * G\right)$ is real-analytic on a neighborhood of α^*. Hence the Hessian of condition (4) is well-defined.

We make the following observations about the notion of generic extremality. When the genericity of $I_n - t^*\Delta$ occurs in numerical examples, it can be rigorously verified. However, in Section 2.6 when we generate counterexamples to the rational dilation conjecture, these counterexamples ultimately will not logically depend upon generic extremality. In practice, random choices of Δ result in generically extremal grammians $I_n - t^*\Delta$. Finally, note that in light of Proposition 2.2.46 genericity would never occur if T_Ω^0 in condition (2) of the definition is replaced by T_Ω. Thus, the reduction from T_Ω to T_Ω^0 provides more than merely a reduction in dimensionality of the optimization problem (2.5.1).

2.5.2. The Second Step, in the Boundary, $\partial \mathcal{S}_z^w$. In Section 2.3, we saw that hyperextremal elements of $\mathcal{S}_z$ possess useful theoretical properties. In particular, in the case $m = 2$ and $n = 4$ we derived estimates for the distance between a nonsingularly approximately hyperextremal grammian Ξ and $\mathcal{D}_z$. Unfortunately, we do not have a direct theoretical construction of hyperextremal grammians. In this section we present a method that generates nonsingularly approximately hyperextremal grammians in the case $m = 2$ and $n = 4$. The method is based on the geometry of the weakly normalized spectral set body, which we now introduce.

Recall from Section 2.2.3 that a grammian G is weakly normalized if $tr(G) = n$. Define the *weakly normalized spectral set body*, $\mathcal{S}_z^w$, by
(2.5.4)
$$\mathcal{S}_z^w = \{G \in \mathbb{C}^{n,n} : G > 0, trG = n, \text{ and } G^{\frac{1}{2}} D_z G^{-\frac{1}{2}} \text{ has } \Omega^- \text{ as a spectral set}\}.$$

Propositions 2.2.38 and 2.2.49 hold for $\mathcal{S}_z^w$ when "weakly normalized" replaces "strongly normalized" in the statements of the propositions. As a corollary, we obtain that $\mathcal{S}_z^w$ is convex. However, unlike $\mathcal{S}_z$, $\mathcal{S}_z^w$ is *not* compact.

Note that $\mathcal{S}_z$ is a subset of $\mathcal{S}_z^w$. In particular, the generically extremal grammian $I_n - t^*\Delta$ obtained in the previous section is also an element of the weakly normalized spectral set body, $\mathcal{S}_z^w$. For particular choices of m and n (e.g. $m = 2$ and $n = 4$), if $I_n - t^*\Delta$ is generically extremal, then there exists a perturbation direction matrix Υ such that if $G(s)$ denotes the ray of grammians

(2.5.5)
$$G(s) = I_n - t^*\Delta + s\Upsilon$$

then for $s > 0$ and sufficiently small

(2.5.6)
$$G(s) \in \partial \mathcal{S}_z^w$$

and

(2.5.7) $$[\gamma(\alpha^*) * G(s)] u^* = 0.$$

Once such an *admissible direction* Υ is determined, we then calculate the *corresponding step*, s^*, i.e. the largest s such that $G(s) \in \partial S_z^w$. This results in a weakly normalized grammian $G(s^*)$. Finally, we obtain a strongly normalized grammian Ξ by "strongly normalizing" $G(s^*)$, i.e. by setting

(2.5.8) $$\Xi = \left[\frac{G(s^*)_{ij}}{\sqrt{G(s^*)_{ii}}\sqrt{G(s^*)_{jj}}} \right]_{i,j=1}^n.$$

Below, we shall describe in greater detail the procedure just outlined for generating hyperextremals. However, we first make several remarks to explain why we found it desirable to use the two normalizations – weak and strong – on the grammians.

Remark 1: Recall that grammian coordinates were introduced to give a theoretically useful parameterization of the diagonalizable $n \times n$ matrices with given eigenvalues $z_1, \ldots, z_n$. Both the strong and weak normalizations possess a certain amount of redundancy. Specifically, let $\cong$ denote unitary equivalence of matrices and let $\sim$ denote Schur equivalence on grammians (i.e. $G_1 \sim G_2$ if and only if there exists a diagonal matrix D such that $G_2 = D^* G_1 D$). Now, if G_1 and G_2 are positive definite matrices, then

(2.5.9) $$G_1^{\frac{1}{2}} D_z G_1^{-\frac{1}{2}} \cong G_2^{\frac{1}{2}} D_z G_2^{-\frac{1}{2}}$$

whenever

(2.5.10) $$G_1 \sim G_2.$$

Consequently, both the weak and strong spectral set bodies are parameterizations of the diagonalizable matrices that have Ω^- as a spectral set, but with the weak body being a more redundant parameterization than the strong. Also, given a grammian $G \in S_z^w$, it is clear that

$$\tilde{G} = \left[\frac{g_{ij}}{\sqrt{g_{ii}}\sqrt{g_{jj}}} \right],$$

i.e. the "strong normalization of G", is in S_z and in fact represents the same diagonalizable matrix as G.

Remark 2: The reason why we computed the first step in S_z rather than S_z^w is because the compactness of S_z implies that if $t^* = \max\{t : I_n - t\Delta \in S_z\}$ as in (2.5.1), then the grammian $I_n - t^*\Delta \in \partial S_z$. With a step in the weakly normalized body S_z^w, because of the greater redundancy, there is no guarantee that the solution of $\sup\{t : I_n - t\Delta \in S_z^w\}$ yields a positive definite grammian.

For example, let $D_{\tilde{z}}$ denote the $(n-1) \times (n-1)$ diagonal matrix whose ii^{th} entry is z_i, and fix $\tilde{G} \in int(S_{\tilde{z}})$. Form the $n \times n$ matrix

$$G = \left(\frac{n}{n-1} \right) \tilde{G} \oplus 0,$$

and note that $trG = n$ but G is not positive definite. Let $\Delta = I_n - G$ and observe that

$$\gamma(\alpha) * (I_n - t\Delta) = \gamma(\alpha) * [(1-t)I_n + tG] > 0 \text{ for } 0 \leq t < 1,$$

so that $I_n - t\Delta \in \mathcal{S}_z^w$ for $0 \leq t < 1$. Hence, $\max\{t : I_n - t\Delta \in \mathcal{S}_z^w\}$ does not exist. Furthermore, if $t = 1$, $I_n - t\Delta = G$ is not positive definite, and hence the solution of $\sup\{t : I_n - t\Delta \in \mathcal{S}_z^w\}$ is 1, and yields a non-positive definite grammian.

Furthermore, this example is stable in the sense that if Δ' is an $n \times n$ matrix with $\|\Delta - \Delta'\|$ small, then the optimization problem $\max\{t : I_n - t\Delta' \in \mathcal{S}_z^w\}$ also has no solution. Finally, note that we cannot fix the problem encountered in this example by renormalizing grammians on the path $\{I_n - t\Delta, 0 \leq t \leq 1\}$. For, if I_k denotes the $k \times k$ identity matrix, and for $0 \leq t < 1$, we set $D_t = \left(\frac{1}{(1-t)+t\frac{n}{n-1}}\right)^{\frac{1}{2}} I_{n-1} \oplus \left(\frac{1}{1-t}\right)^{\frac{1}{2}}$, then

$$D_t^*(I_n - t\Delta)D_t = \left[I_{n-1} + \frac{t\frac{n}{n-1}}{(1-t)+t\frac{n}{n-1}}\left(\tilde{G} - I_{n-1}\right)\right] \oplus 1 \in \mathcal{S}_z$$

$$\xrightarrow{t \to 1} \tilde{G} \oplus 1 \in int(\mathcal{S}_z).$$

Remark 3: In the previous remark we demonstrated a few disadvantages of the greater redundancy in the weak normalization. In this remark, we point out a key benefit. Analysis of the ray $G(s)$ defined in (2.5.5) reveals that the construction of Ξ intrinsically in the strong coordinates would be impractical. Admissible direction would lose their desirable global properties, and the ray $G(s)$ would become an algebraic curve determined locally in the tangent bundle of $\partial \mathcal{S}_z$.

We now describe a method for finding an admissible direction Υ and calculating the corresponding step, s^*.

The following proposition gives sufficient conditions for admissibility.

PROPOSITION 2.5.11. *Assume that $I_n - t^*\Delta$ is generically extremal and Υ is a self-adjoint $n \times n$ matrix satisfying:*

(2.5.12) $$tr(\Upsilon) = 0,$$

(2.5.13) $$[\gamma(\alpha^*) * \Upsilon] u^* = 0, \text{ and}$$

(2.5.14) $$\left\langle \left(\frac{d}{d\alpha_r}\gamma(\alpha)|_{\alpha=\alpha^*} * \Upsilon\right) u^*, u^* \right\rangle = 0, \text{ for } r = 1, \ldots, m.$$

If we define the matrix pencil

(2.5.15) $$G(s) = I_n - t^*\Delta + s\Upsilon,$$

then $G(s) \in \partial \mathcal{S}_z^w$ and $[\gamma(\alpha^*) * G(s)] u^* = 0$ for s positive and sufficiently small.

PROOF. Define a continuous real-valued function F on T_Ω^0 by setting

(2.5.16) $$F(\alpha) = \lambda_{\min}\left[\gamma(\alpha) * (I - t^*\Delta)\right].$$

Since $I_n - t^*\Delta$ is generically extremal, $F(\alpha) \geq 0$ for all $\alpha \in T_\Omega^0$, and there exists a unique $\alpha^* \in T_\Omega^0$ such that $F(\alpha^*) = 0$. Furthermore, there exists a neighborhood

$\mathcal{O} \subseteq T_\Omega^0$ with $\alpha^* \in \mathcal{O}$ such that $F|_\mathcal{O}$ is real analytic and $H_F|_\mathcal{O} > 0$, where H_F denotes the Hessian of F. Also, since T_Ω^0 is compact there exists a $\delta > 0$ such that

$$(2.5.17) \qquad F(\alpha) > \delta \text{ on } T_\Omega^0 \setminus \mathcal{O}.$$

Now define $F_s(\alpha) = \lambda_{\min}\left[\gamma(\alpha) * (I - t^*\Delta) + s\left(\gamma(\alpha) * \Upsilon\right)\right]$. Clearly, since $H_F > 0$ on $\mathcal{O}$, so also $H_{F_s} > 0$ on $\mathcal{O}$ for s sufficiently small. On the other hand, (2.5.14) guarantees that F_s has a critical point at α^*. Since (2.5.13) implies that $F_s(\alpha^*) = 0$, we conclude that
$$(2.5.18)$$
$F_s(\alpha^*) = 0$ and $F_s(\alpha) \geq 0$ for all $\alpha \in \mathcal{O}$ and $s > 0$ and sufficiently small.

But we also see that (2.5.17) implies that

$$(2.5.19) \qquad F_s(\alpha) \geq \delta \text{ for all } \alpha \in T_\Omega \setminus \mathcal{O} \text{ and } s > 0 \text{ and sufficiently small.}$$

With $G(s)$ defined as in (2.5.15), note that (2.5.12) implies that $G(s)$ is weakly normalized for s sufficiently small. Also note that (2.5.18) and (2.5.19) imply that

$$(2.5.20) \qquad \gamma(\alpha) * G(s) \geq 0 \text{ for all } \alpha \in T_\Omega^0 \text{ and } s \text{ sufficiently small,}$$

and

$$(2.5.21) \qquad [\gamma(\alpha^*) * G(s)] u^* = 0.$$

$\square$

Proposition 2.5.11 asserts that if a self-adjoint matrix Υ satisfies (2.5.12), (2.5.13), and (2.5.14), then Υ is an admissible direction. The determining equations (2.5.12), (2.5.13), and (2.5.14) for an admissible direction Υ can be expressed as a homogeneous linear system in the following way. View the $n \times n$ complex self-adjoint matrices as an n^2 dimensional vector space over $\mathbb{R}$, and let $B_1, B_2, \ldots, B_{n^2}$ be a basis. Express the unknown admissible direction Υ as

$$(2.5.22) \qquad \Upsilon = \sum_{k=1}^{n^2} b_k B_k.$$

Setting $\mathbf{b} = (b_1, \ldots, b_{n^2})^t$, and substituting (2.5.22) into (2.5.12), (2.5.13), and (2.5.14) yields a homogeneous linear system

$$(2.5.23) \qquad U\mathbf{b} = 0$$

in the variables $b_1, \ldots, b_{n^2}$. In the case $m = 2$ and $n = 4$ the system $Ub = 0$ is underdetermined, with 6 free variables. For each choice of these free variables, one obtains an admissible direction Υ.

Once an admissible direction Υ is determined, we calculate the step, s^*, corresponding to Υ by solving the following optimization problem:

$$(2.5.24) \qquad s^* = \max_{s \geq 0}\{s : I_n - t^*\Delta + s\Upsilon \in \mathcal{S}_z^w\}.$$

To solve this problem, we use Proposition 2.2.49 and set $G(s) = I_n - t^*\Delta + s\Upsilon$ to write (2.5.24) as

$$s^* = \max_{s \geq 0}\{s : \gamma(\alpha) * G(s) \geq 0 \text{ for all } \alpha \in T_\Omega^0 \}.$$

Next, we use Sylvester's criterion, which states that an $n \times n$ matrix is positive definite if and only if the n leading principle minors are positive. Specifically, for each $k = 1, \ldots, n$ define a $k \times k$ matrix-valued function $P_k(\alpha, s)$ by

$$P_k(\alpha, s) = k \times k \text{ leading principle submatrix of } \gamma(\alpha) * G(s).$$

Since $I_n - t^*\Delta$ is generically extremal, $\gamma(\alpha) * (I - t^*\Delta) > 0$ for all $\alpha \in T_\Omega^0 \setminus \{\alpha^*\}$. Hence $\det P_r(\alpha, 0) > 0$ for $k = 1, \ldots, n$, and $\alpha \neq \alpha^*$. Let $\mathbb{R}^+$ denote the strictly positive reals, and define

$$p_k(\alpha) = \inf(\{s : \det P_r(\alpha, s) < 0\} \cap \mathbb{R}^+),$$

where we make the convention that $\inf(\emptyset) = +\infty$. In particular, note that condition (2.5.13) implies that $p_n(\alpha^*) = +\infty$. Let

$$s(\alpha) = \min\{p_k(\alpha) : k = 1, \ldots, n\}.$$

By Sylvester's criteria, $\gamma(\alpha) * G(s) > 0$ if and only if $s(\alpha) > 0$. Hence,

(2.5.25) $$s^* = \min_{\alpha \in T_\Omega^0} s(\alpha).$$

Just as the step defined in Section 2.5.1 led to grammians $I_n - t^*\Delta$ that are extremal in the sense of Definition 2.3.2, generically we might expect that the above procedure will lead to grammians Ξ that are hyperextremal as in Definition 2.3.5. The authors numerical experience performing these two steps confirms the speculation that with random choices of Δ and corresponding admissible directions Υ, the resulting grammian Ξ is approximately hyperextremal.

2.5.3. Semidefinite Programming Formulation of Step in $\mathcal{D}_z$. In Section 2.4, we introduced the step (2.4.1) in the dilation body. In this section, we show how to express (2.4.1) as a semidefinite program. Using the condition (2.2.41) from Proposition 2.2.40 for $I_n - \tau(I_n - \Xi)$ to be an element of $\mathcal{D}_z$ leads to the following problem:

(2.5.26) $$\max \tau$$

subject to

(2.5.27) $$h(\lambda) * (I_n - \tau(I_n - \Xi))) + \sum_{r=1}^{m} \varphi_r(\lambda) X_r \geq 0 \text{ for all } \lambda \in \partial\Omega,$$

(2.5.28) $$X_r^* = X_r \text{ for } r = 1, \ldots, m,$$

where the X_r are complex $n \times n$ matrix unknowns. For each fixed λ, (2.5.27) is a linear matrix inequality. Thus, the above problem is a continuously parameterized system of linear matrix inequalities. We approximate the problem (2.5.26)-(2.5.28) by choosing a a finite set $\mathcal{P} = \{\lambda_1, \ldots, \lambda_L\} \subseteq \partial\Omega$ and evaluating condition (2.5.27) at each element of $\mathcal{P}$. We also expand the matrix unknowns X_r in terms of a basis $\{B_k\}_{k=1,\ldots n^2}$ for the complex $n \times n$ self-adjoint matrices, by setting

$$X_r = \sum_{k=1}^{n^2} x_{r,k} B_k.$$

We also define the following vectors in $\mathbb{R}^{mn^2+1}$:

(2.5.29) $$\mathbf{x} = \left(\tau, x_{1,1}, \ldots, x_{1,n^2}, x_{2,1}, \ldots, x_{2,n^2}, \ldots, x_{m,1}, \ldots, x_{m,n^2}\right)^t,$$

(2.5.30) $$\mathbf{c} = (1, 0, 0, \ldots, 0)^t.$$

The problem (2.5.26)-(2.5.28) is approximated by the following semidefinite program: written in the form:

(2.5.31) $$\max \mathbf{c}^t \mathbf{x}$$

subject to

(2.5.32)
$$h(\lambda_l) * I_n + x_1\left(h(\lambda_l) * (\Xi - I_n)\right) + \sum_{r=1}^{m} \sum_{k=1}^{n^2} x_{r,k} \left(\varphi_r(\lambda_l) B_k\right) \geq 0, 1 \leq l \leq L.$$

We solve this problem using standard algorithms of semidefinite programming.

As a consequence Proposition 2.4.36, we need not undertake a careful examination of the algorithms that we employ to solve the problem (2.4.1). Rather, if we have a grammian $I_n - \tau^* (I_n - \Xi)$ with $\tau^* < 1$ and self-adjoint matrices $X_1, \ldots, X_m$ that satisfy the hypothesis of Proposition 2.4.36, then the proposition yields a lower bound for the distance between Ξ and the dilation body, $\mathcal{D}_z$. In Section 2.6 we apply this distance estimate to the nonsingularly hyperextremal grammian Ξ, constructed in Section 2.5.2, and obtain a second verification that Ξ lies outside $\mathcal{D}_z$.

2.5.4. Computation of the Functions. The algorithm that we use to compute the functions φ_r, w, H_λ, and Γ_α is described in Chapter 3, and we now briefly summarize the main points of this algorithm. The key feature of this algorithm is that it permits evaluation of these functions with very high levels of precision.

We assume that the boundary $\partial \Omega = \partial_0 \cup \cdots \cup \partial_m$ of the domain Ω consists of $m + 1$ circles, where ∂_0 is the unit circle, and for $r = 1, \ldots, m$, ∂_r is a circle of radius $\rho_r < 1$ and center c_r with $|c_r| < 1$. As a result of Proposition 2.2.1, the function $w(\alpha)$ is directly computable from the functions $\varphi_1, \ldots, \varphi_m$. Furthermore, the functional identity (2.2.10) allows us to compute the values of the function Γ_α directly from the values of H_λ. We now describe the computation of the functions $\varphi_1, \ldots, \varphi_m$ and H_λ.

Recall from Section 2.2.1 that $\sigma = \sigma_0 + \cdots + \sigma_m$, where σ_r denotes normalized arc length measure on ∂_r. Let $L^{2,h}(d\sigma)$ be the closure in $L^2(d\sigma)$ of $ReRat(\Omega^-)|_{\partial \Omega}$. For $r = 0, \ldots, m$ and $l \geq 1$, define rational functions

$$F_{r,l}(z) = \begin{cases} z^l, & \text{if } r = 0, \\ \left(\frac{\rho_r}{z - c_r}\right)^l, & \text{if } r \geq 1 \end{cases}$$

and for all $r = 0, \ldots, m$ and $l \geq 1$ define real-valued rational functions

(2.5.33) $$f_{r,2l-1}(z) = ReF_{r,l}$$
(2.5.34) $$f_{r,2l}(z) = ImF_{r,l}.$$

Additionally we define the constant function, $f_{0,0}(z) \equiv \frac{1}{\sqrt{2(m+1)}}$ on $\partial \Omega$.

Evidently, Runge's Theorem [**Rud87**] implies that the functions $f_{r,n}(z)$ have dense linear span in $L^2(d\sigma)$. We enumerate the set of functions $\{f_{r,k} : r =$

$0, \ldots, m, k \geq 1\}$ by fixing an $m+1$ tuple of positive even integers $\mathbf{N} = (N_0, \ldots, N_m)$, letting
$$N = N_0 + N_1 + \cdots + N_m,$$
and ordering the functions $f_{r,k}$ into "blocks of size N" as follows. Set $f_1, \ldots, f_N$ equal to
$$f_{0,1}, \ldots, f_{0,N_0}, f_{1,1}, \ldots, f_{1,N_1}, \ldots, f_{m,1}, \ldots, f_{m,N_m},$$
respectively, and set $f_{N+1}, \ldots, f_{2N}$ equal to
$$f_{0,N_0+1}, \ldots, f_{0,2N_0}, f_{1,N_1+1}, \ldots, f_{1,2N_1}, \ldots, f_{m,N_m+1}, \ldots, f_{m,2N_m},$$
respectively, etc. Let $f_0 = f_{0,0} = \frac{1}{\sqrt{2(m+1)}}$. Let $\mathbf{f}$ denote the infinite column vector
$$\mathbf{f} = (f_0, f_1, \ldots, f_k, \ldots)^t$$
where t denotes the transpose, and for each positive integer n, let $\mathbf{f_n}$ denote the finite column vector
$$\mathbf{f_n} = (f_0, f_1, \ldots, f_n)^t.$$
Let $\mathcal{M}_n$ denote the linear span of the functions $f_0, \ldots, f_n$ in $L^{2,h}(d\sigma)$, and let $P_{\mathcal{M}_n}$ denote the orthogonal projection of $L^{2,h}(d\sigma)$ onto $\mathcal{M}_n$.

Given a function $g \in L^{2,h}(d\sigma)$, we wish to evolve a formula for calculating $P_{\mathcal{M}_n} g$ in terms on the functions $f_0, \ldots, f_n$. To achieve this end, let $\mathbf{u_n} = (u_0, u_1, \ldots, u_n)$ be the vector of functions that results from Gram-Schmidt orthogonalization applied to the sequence $f_0, f_1, \ldots, f_n$, let G_n be the $(n+1) \times (n+1)$ grammian matrix $G_n = [\langle f_j, f_i \rangle]_{i,j=0}^n$, and let $G_n^{-1} = C^t C$ be the Cholesky decomposition of G_n^{-1}. Evidently, with these notations, we have that $\mathbf{u_n} = C\mathbf{f_n}$. Hence, if $g \in L^{2,h}(d\sigma)$ then

$$(2.5.35) \quad P_{\mathcal{M}_n} g = \sum_{k=0}^n \langle g, u_k \rangle u_k = \sum_{k=1}^{n+1} \langle g, (C\mathbf{f_n})_k \rangle (C\mathbf{f_n})_k = \langle g, \mathbf{f_n} \rangle G_n^{-1} \mathbf{f_n},$$

where $\langle g, \mathbf{f_n} \rangle$ denotes the row vector $(\langle g, f_0 \rangle, \langle g, f_1 \rangle, \ldots, \langle g, f_n \rangle)$.

We can compute the inner products $\langle f_j, f_i \rangle$, the entries of G_n, exactly using residue theory.

Now $L^{2,h}(d\sigma)$ has codimension m in $L^2(d\sigma)$ and the functions $\varphi_1, \ldots, \varphi_m$ form an orthonormal basis for the orthogonal complement of $L^{2,h}(d\sigma)$ in $L^2(d\sigma)$. For $r = 1, \ldots, m$, let χ_r be the harmonic measure of ∂_r, i.e.

$$\chi_r(z) = \begin{cases} 1, z \in \partial_r \\ 0, \text{ otherwise.} \end{cases}$$

The functions $\chi_1, \ldots, \chi_m$ are linearly independent modulo $L^{2,h}(d\sigma)$. Let $\psi_r = P_{L^{2,h}(d\sigma)^\perp} \varphi_r$, so that necessarily $\psi_1, \ldots, \psi_m$ span $(L^{2,h}(d\sigma))^\perp$. Then we obtain a formula for $\varphi_1, \ldots, \varphi_m$ by applying Gram-Schmidt orthogonalization to the functions $\psi_1, \ldots, \psi_m$.

To obtain a formula for $H_\lambda(z)$ we employ the following "pole-matching" technique. For $z \in \partial\Omega$ and $\lambda \in \partial_0$ let

$$p_\lambda^{D_0}(z) = \text{Re} \left[\frac{\lambda + z}{\lambda - z} \right]$$

be the Poisson kernel for the unit disk D_0, and set $P_\lambda^{D_0} = \frac{\lambda+z}{\lambda-z}$. For $r = 1, \ldots, m$ let

$$p_\lambda^{D_r}(z) = Re\left[\frac{\lambda + z - 2c_r}{z - \lambda}\right]$$

be the Poisson kernel for the disk D_r with boundary ∂_r and containing the point at ∞. Define $P_\lambda^{D_r}$, a holomorphic function on Ω whose real part is $p_\lambda^{D_r}$ by

$$P_\lambda^{D_r}(z) = \frac{\lambda + z - 2c_r}{z - \lambda}.$$

Fix $\lambda \in \partial_r$ and define

$$Q_\lambda^r(z) = \begin{cases} 0, & \text{if } z \in \partial_r \\ p_\lambda^{D_r}, & \text{if } z \in \partial_s, s \neq r. \end{cases}$$

Noting that $Q_\lambda^r(z) = p_\lambda^{D_r}(z) - P_\lambda(z)$ and using (2.2.4) we obtain

$$ReH_\lambda(z) = P_\lambda(z) - \sum_{s=1}^m \varphi_s(\lambda)\varphi_s^\wedge(z)$$

$$= p_\lambda^{D_r}(z) - \left(Q_\lambda^r(z) + \sum_{s=1}^m \varphi_s(\lambda)\varphi_s^\wedge(z)\right)$$

Hence

$$Q_\lambda^r(z) + \sum_{s=1}^m \varphi_s(\lambda)\varphi_s^\wedge(z) = p_\lambda^{D_r}(z) - ReH_\lambda(z)$$

is a harmonic function on Ω with single-valued harmonic conjugate. Since $\varphi_r \in \left(L^{2,h}(d\sigma)\right)^\perp$, we have

$$Q_\lambda^r(z) + \sum_{s=1}^m \varphi_s(\lambda)\varphi_s^\wedge(z) = P_{L^{2,h}(d\sigma)}\left[Q_\lambda^r(z) + \sum_{s=1}^m \varphi_s(\lambda)\varphi_s^\wedge(z)\right]$$

$$= P_{L^{2,h}(d\sigma)}Q_\lambda^r(z)$$

We obtain the following formula for ReH_λ:

(2.5.36) $$ReH_\lambda(z) = p_\lambda^{D_r}(z) - P_{L^{2,h}(d\sigma)}Q_\lambda^r(z).$$

To obtain a formula for H_λ we define the following holomorphic functions. According to the definition of our basis given in (2.5.33) and (2.5.34), if f_k is an element of our basis then for all $k \geq 1$, f_{2k} is a harmonic conjugate of f_{2k-1} and $-f_{2k-1}$ is a harmonic conjugate of f_{2k}. For each $k \geq 1$, we define a holomorphic function $\hat{F}_k$ on Ω by

(2.5.37) $$\hat{F}_0 = \frac{1}{\sqrt{2m+1}}$$

(2.5.38) $$\hat{F}_{2k-1} = f_{2k-1} + if_{2k}$$

(2.5.39) $$\hat{F}_{2k} = f_{2k} - if_{2k-1}.$$

Thus, for all $j \geq 0$, $f_j = Re\hat{F}_j$. Equivalently, with these notations, and according to the indexing procedure described for the functions f_k, the functions $\hat{F}_k$ are merely the partition into blocks of size N of the functions $\hat{F}_{r,k}$ defined for $r = 0, \ldots, m$ and $l \geq 1$ by

(2.5.40) $$\hat{F}_{0,0} = \frac{1}{\sqrt{2m+1}}$$

(2.5.41) $$\hat{F}_{r,2l-1} = F_{r,l}$$

(2.5.42) $$\hat{F}_{r,2l} = -iF_{r,l}.$$

Let $\mathbf{F_N} = (\hat{F}_0, \ldots, \hat{F}_N)^t$. Define the row vector of functions on $\partial\Omega$ by
$$\mathbf{q_N}(\lambda) = (\langle Q_\lambda^r, f_0 \rangle, \langle Q_\lambda^r, f_1 \rangle, \ldots, \langle Q_\lambda^r, f_N \rangle) \text{ if } \lambda \in \partial_r.$$

Using residue theory, the vector $\mathbf{q_N}$ can be computed exactly as a rational function of λ. By applying equation (2.5.35), and adding harmonic conjugates to (2.5.36) we obtain the following formula for H_λ
$$H_\lambda(z) \approx P_\lambda^{D_r}(z) - \mathbf{q_N}(\lambda) G_N^{-1} \mathbf{F_N}(z).$$

Further details about the convergence properties and error of this approximation can be found in Chapter 3.

2.5.5. Solving Optimization Problems. Our approach to the rational dilation conjecture also requires the solution of several optimization problems of the form

(2.5.43) $$\min_{x \in \mathcal{D}} h(x)$$

where $\mathcal{D}$ is either T_Ω^0 or $\partial\Omega$ and h is a piecewise C^2 function on $\mathcal{D}$. In particular, problems (2.5.2) for the first step, (2.5.25) for the second step, (2.4.11) for C^p, and (2.4.12) for C^h are of this form.

We solve these optimization problems by the following two-step procedure. First, select a finite set of points $\mathcal{P} \subseteq \mathcal{D}$ and compute
$$\xi_0 = \min_{x \in \mathcal{P}} h(x).$$

Second, use ξ_0 as the initial value for a BFGS quasi-Newton method [**GMW81**] and compute

(2.5.44) $$\xi^* = \min_{x \in \mathcal{D}} h(x).$$

In cases where we lack a formula for h', we use central difference approximations.

We now describe a second, qualitatively different role that the optimization problem (2.5.43) will play. To justify the counterexample in Section 2.6, we shall need to establish inequalities (2.2.50), (2.3.44), and (2.4.35). If we let $M(x)$ denote a self-adjoint matrix-valued function over $\mathcal{D}$, then these inequalities are all of the form
$$M(x) \geq 10^{-N}, \text{ for all } x \in \mathcal{D},$$
or equivalently,

(2.5.45) $$\lambda_{\min}[M(x)] \geq 10^{-N}, \text{ for all } x \in \mathcal{D},$$

where, as before, $\mathcal{D}$ is either T_Ω^0 or $\partial\Omega$.

Letting $h(x) = \lambda_{\min}[M(x)]$ and letting ξ^* be as in (2.5.44), it is clear that (2.5.45) holds if and only if
$$\xi^* \geq 10^{-N}.$$

Of course, we cannot calculate ξ^* exactly on a computer. Rather, the algorithm just described yields as approximation, $\hat{\xi}^*$, to the true minimum ξ^*. Nevertheless, we see that if $\hat{\xi}^*$ is accurate to within an order of magnitude then (2.5.45) will hold provided $\hat{\xi}^* \geq 10^{-N}$. We formalize this relationship between the accuracy in the computation of ξ^* and the inequality (2.5.45) in the following definition.

DEFINITION 2.5.46. *Let $M(x)$ be a self-adjoint matrix-valued function on $\mathcal{D}$, and let $\hat{\xi}^*$ be calculated via the two-step optimization procedure outlined above. $\hat{\xi}^* \in \mathbb{R}$ is a **sufficiently accurate minimum** provided*

$$\min_{x \in \mathcal{D}} \lambda_{\min}[M(x)] \geq 10^{\lfloor \log_{10} \hat{\xi}^* \rfloor}.$$

2.6. A Computational Counterexample

In this section, we apply the procedure described in Sections 2.5.1 and 2.5.2 to obtain a concrete grammian Ξ^* that is nonsingularly approximately hyperextremal. We then employ the results of Sections 2.3 and 2.4 to see that – subject to certain assumptions about the accuracy of our computation – Ξ^* gives rise to a counterexample to the rational dilation conjecture.

We first give our choices for the objects required to implement the procedure of Sections 2.5.1 and 2.5.2. Let Ω be the two-holed domain whose boundary consists of the circles $\partial_0, \partial_1, \partial_2$, where ∂_0 is the unit circle, ∂_1 has center at $z = 0$ and radius $\frac{1}{10}$, and ∂_2 has center at $z = \frac{1}{2}$ and radius $\frac{1}{4}$. We fix the point $\lambda^* = e^{\frac{381\pi i}{500}}$. We compute the functions φ_r, Γ_α, and H_λ according to the formulae described in Section 2.5.4 with $\mathbf{N} = (120, 60, 120)$. With this choice of $\mathbf{N}$, we expect that the functions are accurate to 10^{-16}. We also fix $\mathbf{N_0} = (240, 240, 240)$.

Fix the eigenvalues $z_1 = -\frac{3}{8} + \frac{i}{2}$, $z_2 = \frac{1}{5} + \frac{23i}{29}$, $z_3 = \frac{9}{16} - \frac{i}{2}$, $z_4 = -\frac{1}{8} + \frac{15i}{37}$. Figure 2.2 shows the domain Ω and the spectral points z_i. Define the step direction, Δ, by first setting

$$\tilde{\Delta} = \begin{pmatrix} 0 & \frac{1}{3} & -\left(\frac{5}{8}\right) & \frac{5}{3} \\ \frac{1}{3} & 0 & \frac{i}{6} & -\left(\frac{12}{37}\right) - 2i \\ -\left(\frac{5}{8}\right) & \frac{-i}{6} & 0 & \frac{23}{31} + \frac{3i}{2} \\ \frac{5}{3} & -\left(\frac{12}{37}\right) + 2i & \frac{23}{31} - \frac{3i}{2} & 0 \end{pmatrix},$$

and then setting

$$\Delta = \frac{\tilde{\Delta}}{\|\tilde{\Delta}\|}.$$

Implementing the first step in the direction Δ as described in Section 2.5.1, and computing the step t^* via (2.5.2) yields

$$t^* \approx 0.1491104563.$$

Using the conditions in Proposition 2.5.11, we compute an admissible direction $\tilde{\Upsilon}$ and normalize the resulting matrix to obtain the direction

$$\Upsilon_\mathbf{N} = [v_{ij}]_{i,j=1}^4,$$

where

$$v_{11} = -0.5811389 \qquad v_{12} = 0.0349211 + 0.6450892\,i$$
$$v_{13} = -0.00568712 + 0.01819725\,i \qquad v_{14} = -0.0010246357 + 0.0005745767\,i$$
$$v_{22} = -0.002376318 \qquad v_{23} = -0.005559324 - 0.010834760\,i$$
$$v_{24} = 0.0009504201 - 0.0009751565\,i \qquad v_{33} = 0.2857078$$
$$v_{34} = -0.00886752 - 0.02057074\,i \qquad v_{44} = 0.2978074$$

2.6. A COMPUTATIONAL COUNTEREXAMPLE

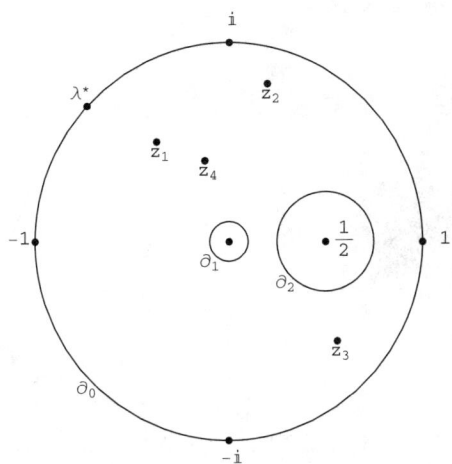

FIGURE 2.2. The domain Ω and spectral points z_i.

and the entries below the main diagonal are determined by the fact that $\Upsilon_{\mathbf{N}}$ is self-adjoint. With this choice of Υ, we compute the corresponding step s^* according to (2.5.25), obtaining
$$s^* = 0.6213862975.$$
Finally, following (2.5.5) we set
$$G(s^*) = I - t^*\Delta + s^*\Upsilon,$$
which we strongly normalize via (2.5.8) to obtain the matrix
$$\Xi^* = [\xi_{ij}]_{i,j=1}^4,$$
where

$\xi_{11} = 1.000000000$ $\xi_{12} = 0.5019369590$
$\xi_{13} = 0.03135211909$ $\xi_{14} = 0.08735647323$
$\xi_{22} = 1.000000000$ $\xi_{23} = 0.010503874251 - 0.008544191161\,i$,
$\xi_{24} = 0.08241925793 - 0.01500227467\,i$ $\xi_{33} = 1.000000000$
$\xi_{34} = 0.00134013061 + 0.07582405338\,i$ $\xi_{44} = 1.000000000$

and the entries below the main diagonal are determined by the fact that Ξ^* is self-adjoint. We now show that Ξ^* gives rise to a counterexample to the rational dilation conjecture. This result is based on the following two claims about Ξ^*.

CLAIM 1. $\min_{G \in \mathcal{S}_z} \|\Xi^* - G\| < 10^{-13}$, i.e. Ξ^* within 10^{-13} of $\mathcal{S}_z$.

JUSTIFICATION. To justify this claim, let I denote the 4×4 identity matrix, and let
$$\Xi_t^* = I - t \frac{(I - \Xi^*)}{\|I - \Xi^*\|}.$$
Note that $\Xi_{\|I-\Xi^*\|}^* = \Xi^*$ and $\|\Xi^* - \Xi_{\|I-\Xi^*\|-10^{-13}}^*\| = 10^{-13}$. Figure 2.3 shows the graphs of the functions $\lambda_{\min} \left[\gamma^{\mathbf{N}}(\alpha) * \Xi_{\|I-\Xi^*\|-10^{-13}}^* \right]$ for $\alpha \in T_\Omega^0$ and
$$\min_{\alpha_2 \in \partial_2} \left[\lambda_{\min} \left[\gamma_{(\lambda^*, \alpha_1, \alpha_2)}^{\mathbf{N}} * \Xi_{\|I-\Xi^*\|-10^{-13}}^* \right] \right].$$

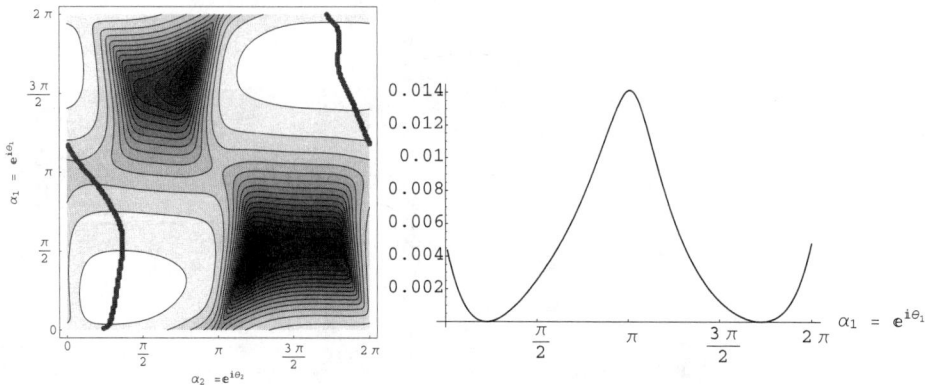

FIGURE 2.3. Graphs of $\lambda_{\min}\left[\gamma^{\mathbf{N}}(\alpha) * \Xi^*_{\|I-\Xi^*\|-10^{-13}}\right]$ (left) and $\min_{\alpha_2 \in \partial_2}\left[\lambda_{\min}\left[\gamma^{\mathbf{N}}_{(\lambda^*,\alpha_1,\alpha_2)} * \Xi^*_{\|I-\Xi^*\|-10^{-13}}\right]\right]$ (right), where the segment in the left plot indicates $\arg\min_{\alpha_2 \in \partial_2} \lambda_{\min}\left[\gamma^{\mathbf{N}}(\alpha) * \Xi^*_{\|I-\Xi^*\|-10^{-13}}\right]$

The function $\lambda_{\min}\left[\gamma^{\mathbf{N}}(\alpha) * \Xi^*_{\|I-\Xi^*\|-10^{-13}}\right]$ has two local minima at

$$\alpha^* = \{-0.7289686274 + 0.6845471059\,i, 0.07826975614 + 0.06224022232\,i,$$
$$0.6414295111 + 0.2061496869\,i\}$$

and

$$\beta^* = \{-0.7289686274 + 0.6845471059\,i, 0.06262423551 - 0.07796284452\,i,$$
$$0.6943571465 - 0.1572428046\,i\}$$

The corresponding eigenvectors are

$$u^* = \{-0.010543001468 - 0.004188041820\,i, -0.00023219063 - 0.03593872566\,i,$$
$$0.9967956820, -0.05851502476 + 0.03942105650\,i\}$$

and

$$v^* = \{-0.2594575983 + 0.0484685068\,i, 0.9643742646,$$
$$0.002883174371 + 0.007304281540\,i, -0.01322590905 + 0.00884570050\,i\}.$$

We compute that

$$\lambda_{\min}\left(\gamma^{\mathbf{N}}(\alpha^*) * \Xi^*_{\|I-\Xi^*\|-10^{-13}}\right) = 1.775287757 \times 10^{-15}$$

and

$$\lambda_{\min}\left(\gamma^{\mathbf{N}}(\beta^*) * \Xi^*_{\|I-\Xi^*\|-10^{-13}}\right) = 5.011381856 \times 10^{-14}.$$

If we assume that the values given above are sufficiently accurate minima for $\gamma(\alpha) * \Xi^*_{\|I-\Xi^*\|-10^{-13}}$ (according to Definition 2.5.46), then we conclude that $\Xi^*_{\|I-\Xi^*\|-10^{-13}}$ is in $\mathcal{S}_z$. □

2.6. A COMPUTATIONAL COUNTEREXAMPLE

CLAIM 2. $\min_{G \in \mathcal{D}_z} \|\Xi^* - G\| \gg 10^{-9}$, i.e. Ξ^* is significantly more than 10^{-9} outside $\mathcal{D}_z$.

We shall give two independent justifications of this claim: the first employs the results of Section 2.3, and the second employs the results of Section 2.4.

FIRST JUSTIFICATION. The first justification relies on the empirical fact that Ξ^* is a nonsingularly approximately hyperextremal grammian as defined in Definition 2.3.48. Indeed, we compute that

$$\min_{\alpha \in T_\Omega^0} \lambda_{\min}\left[\gamma^{\mathbf{No}}(\alpha) * \Xi^*\right] \approx -2.067629 \times 10^{-16}.$$

Also, if we let

$$\alpha^* = \{-0.7289686274 + 0.6845471059\,i, 0.06262423551 - 0.07796284452\,i$$
$$, 0.6943571465 - 0.1572428046\,i\}$$
$$\beta^* = \{-0.7289686274 + 0.6845471059\,i, 0.06262423551 - 0.07796284452\,i$$
$$, 0.6943571465 - 0.1572428046\,i\}$$
$$u^* = \{-0.010543001468 - 0.004188041820\,i, -0.00023219063 - 0.03593872566\,i,$$
$$0.9967956820, -0.05851502476 + 0.03942105650\,i\}$$
$$v^* = \{-0.2594575983 + 0.0484685068\,i, 0.9643742646,$$
$$0.002883174371 + 0.007304281541\,i, -0.01322590905 + 0.00884570050\,i\}$$

then we have

$$\langle (\gamma^{\mathbf{No}}(\alpha^*) * \Xi^*) u^*, u^* \rangle \approx 3.185598 \times 10^{-18}$$
$$\langle (\gamma^{\mathbf{No}}(\beta^*) * \Xi^*) v^*, v^* \rangle \approx -2.067629 \times 10^{-16}.$$

If we assume that these values are sufficiently accurate minima for $\gamma(\alpha)*\Xi^*+10^{-17}$ then Ξ^* is approximately hyperextremal with parameter $\boldsymbol{\theta} = (\alpha^*, \beta^*, u^*, v^*, 10^{-17})$. We now wish to show that Ξ^* is significantly outside $\mathcal{D}_z$. Using Proposition 2.3.55, we will see that Ξ^* is at least 10^{-11} away from $\mathcal{D}_z$ ((2.6.1) below). Using Proposition 2.3.59, we will see that Ξ^* is at least 10^{-9} away from $\mathcal{D}_z$ ((2.6.3) below). To this end, we fix $\lambda_1 = -\left(\frac{1}{10}\right)$ and $\lambda_2 = \frac{1}{2} - \frac{i}{4}$ and consider the linear system $\mathbf{X} A_{\Xi, \Theta, \lambda_1, \lambda_2} = B_{\Xi, \Theta, \lambda_1, \lambda_2}$ of (2.3.47). Solving this system, we find X_1 and X_2 with

$$\|Im X_1\|_2 \approx 0.002275598559$$

and

$$\|Im X_2\|_2 \approx 0.003189078682.$$

Now let $\epsilon = 10^{-11}$, so that by (2.3.52), $\|\mathbf{E}^\Theta(\epsilon)\| \approx 0.00008242757613$. Let C_1 be the first 8×4 block column of $A_{\Xi, \Theta, \lambda_1, \lambda_2}^{-1}$. Since $\|C_1\| \approx 10.67092873$, we see that

$$\|Im X_1\|_2 = 0.002275598559 > 0.0008795787901 \approx \|C_1\|\|\mathbf{E}^\Theta(\epsilon)\|.$$

Hence, Proposition 2.3.55 implies that there does not exist $G \in \mathcal{D}_z$ with

(2.6.1) $$\|G - \Xi^*\| < 10^{-11}.$$

We also compute

$$\kappa_{X_1}(u^*, v^*) = -0.001254151086 + 0.000807487121\,i$$

and
$$\kappa_{X_2}(u^*, v^*) = 0.002642290606 - 0.001701242904\,i$$
Let $\epsilon = 10^{-9}$. Since

(2.6.2) $\quad |\kappa_{X_1}(u^*, v^*)| = 0.001491620058 > 0.001313212357$
$$= \|\nu_1(\alpha^*)\| \left(E_1^\Theta(\epsilon)^2 + E_2^\Theta(\epsilon)^2\right)^{\frac{1}{2}} + \|\nu_1(\beta^*)\| \left(E_5^\Theta(\epsilon)^2 + E_6^\Theta(\epsilon)^2\right)^{\frac{1}{2}},$$

Proposition 2.3.59 implies that there does not exist $G \in \mathcal{D}_z$ with

(2.6.3) $\quad\quad\quad\quad\quad\quad\quad\quad \|G - \Xi^*\| < 10^{-9}.$

Hence, Ξ^* is in fact further outside $\mathcal{D}_z$, then the estimate given in (2.6.1). □

SECOND JUSTIFICATION. The second justification of Claim 2 uses the results of Section 2.4 which do not require any assumptions about the hyperextremality of Ξ^*. We solve the semidefinite program (2.5.31)-(2.5.32) with $\mathcal{P} = \mathcal{P}_{100}(\partial\Omega)$ and $\Delta = I - \Xi^*$. The solution yields $\tau_{\mathbf{N}}^* \approx 0.9999677305$, and self-adjoint matrices $X_1^{\mathrm{SDP}} = \sum_{k=1}^{n^2} y_{1,k} B_k$, $X_2^{\mathrm{SDP}} = \sum_{k=1}^{n^2} y_{2,k} B_k$. We find that for all $\lambda \in \partial\Omega$

$$\lambda_{\min}\left[h^{\mathbf{N_0}}(\lambda) * \left(I - \tau_{\mathbf{N_0}}^*(I - \Xi^*)\right) + \varphi_1^{\mathbf{N_0}}(\lambda)X_1^{\mathrm{SDP}} + \varphi_2^{\mathbf{N_0}}(\lambda)X_2^{\mathrm{SDP}}\right]$$
$$\leq 2.703066410 \times 10^{-14}.$$

If we assume that 10^{-13} minus the above value is a sufficiently accurate minimum for $10^{-13} - \left(h(\lambda) * (I - \tau_{\mathbf{N}}^*(I - \Xi^*)) + \varphi_1(\lambda)X_1^{\mathrm{SDP}} + \varphi_2(\lambda)X_2^{\mathrm{SDP}}\right)$, then $I - \tau_{\mathbf{N}}^*(I - \Xi^*)$ is approximately dilation extremal with parameter $\boldsymbol{\xi} = (\xi_1, 10^{-13})$ (according to Definition 2.4.33) for ξ_1 sufficiently large. As a consequence Proposition 2.4.36, we need not undertake a careful examination of the algorithms that we employ to solve the semidefinite program (2.5.31)-(2.5.32). Rather, since we have a grammian $I_n - \tau_{\mathbf{N}}^*(I_4 - \Xi^*)$ with $\tau_{\mathbf{N}}^* < 1$ and self-adjoint matrices $X_1^{\mathrm{SDP}}, \ldots, X_m^{\mathrm{SDP}}$ such that $I_4 - \tau_{\mathbf{N}}^*(I_4 - \Xi^*)$ is approximately hyperextremal, then Proposition 2.4.36 yields a lower bound for the distance between Ξ^* and the dilation body, $\mathcal{D}_z$. To apply Proposition 2.4.36, we must estimate the constants C^p and C^{radius} defined in (2.4.11) and (2.4.26), respectively. We compute C^p by applying the two-step optimization procedure described in Section 2.5.5. We obtain a lower bound for C^{radius} by applying the step in $\mathcal{D}_z$ algorithm with $\Delta_k = \pm B_k$ for the matrices $B_1, B_2, \ldots, B_{12}$ that form a basis for the subspace of 4×4 self-adjoint matrices Δ with $\Delta_{ii} = 0$ for $i = 1, \ldots, m$. We obtain the values $\tau_{\mathbf{N}}^*(k)$, and thus the ball of radius $\frac{1}{\sqrt{12}} \min_k \tau_{\mathbf{N}}^*(k)$ is in $\mathcal{D}_z$. We obtain

$$C^p \approx 0.0002007875868$$
$$C^{\mathrm{radius}} \approx 0.009476290830.$$

Let $\epsilon = 10^{-7} < 3.058045824 \times 10^{-7} \approx \frac{1-\tau_{\mathbf{N}}^*}{\tau_{\mathbf{N}}^*} C^{\mathrm{radius}}$. Define δ and ρ according to (2.4.37) and (2.4.38), respectively. Let

$$\eta = \frac{\delta}{\rho + \delta} C^p - \xi_2 \approx 7.767138627 \times 10^{-11}.$$

We consider the system of linear inequalities
$$L_{\lambda_t, x_{\lambda_l}}(\mathbf{Z}) \geq \eta, \text{ for } \lambda_l \in \mathcal{P}_{200}(\partial\Omega).$$

We find that this system is inconsistent by using standard algorithms of linear programming, as implement in Mathematica V4 [**Wol99**] and lpsolve [**Ber**]. By Proposition 2.4.36, we conclude that there does not exist $G \in \mathcal{D}_z$ with

(2.6.4) $$\|G - \Xi^*\| < 10^{-7}.$$

$\square$

In summary, we have shown that the grammian Ξ^* is within 10^{-13} of $\mathcal{S}_z$, but Ξ^* is at least 10^{-7} outside $\mathcal{D}_z$. We conclude that there exists a grammian $\Xi \in \mathcal{S}_z$ with $\|\Xi - \Xi^*\| < 10^{-13}$ small, such that $\Xi \notin \mathcal{D}_z$, thereby obtaining a counterexample to the rational dilation conjecture.

2.7. Plausibility Arguments

In the previous section, the demonstration that the grammian Ξ is a counterexample to the rational dilation question relied on certain machine computations, which we invite others to reproduce. In this section, we present further evidence that our implementation of the machine computations is valid. At the same time, we gain some additional insight into the nature of the counterexample Ξ^*.

First, we perform the computation with increasing values of **N**. Table 2.2 gives the results. Observe that $|t^*_{\mathbf{N}_k} - t^*_{\mathbf{N}_{k-1}}| \to 0$ exponentially fast as $|\mathbf{N}| \to \infty$. However, $\|ImX_r^{\mathbf{N}}\| > 10^{-3}$ and $1 - \tau_{\mathbf{N}}^* > 10^{-5}$ for all **N**.

Using the same algorithms, we also check the following other cases where the dilation conjecture is known to be true.

(1) Ω is an annulus.
(2) Ω is the two-holed domain from the previous section, and M is 2×2 with eigenvalues z_1 and z_2.

The results of these computations are described in [**Rap02**].

Using the algorithms described in this memoir, we gain further insight into the relationship between the sets $\mathcal{S}_z$ and $\mathcal{D}_z$ in a neighborhood of the grammian Ξ^* that we computed in the previous section.

The grammian Ξ^* was obtained by strongly normalizing the grammian $G(s^*) = I_n - t^*\Delta + s^*\Upsilon$. Now, $G(s^*)$ can be viewed as the endpoint of the segment of grammians in $\mathcal{S}_z^w$ defined for $s \in [0, s^*]$ by

(2.7.1) $$G(s) = I_4 - t^*\Delta + s\Upsilon.$$

Recall that $\mathcal{S}_z \subseteq \mathcal{S}_z^w$ and $\mathcal{S}_z^w$ is convex. By construction, $G(0) \in \mathcal{S}_z^w$ and $G(s^*) \in \mathcal{S}_z^w$, and thus $G(s) \in \mathcal{S}_z^w$ for all $s \in [0, s^*]$. We examine the weakly normalized spectral set body $\mathcal{S}_z^w$ and dilation body $\mathcal{D}_z^w$ in a neighborhood of $G(s^*)$; specifically we define for $q \geq 0$

$$\Delta_q = I_4 - G(q).$$

We apply the algorithms for a step in $\mathcal{S}_z^w$ and a step in $\mathcal{D}_z^w$ with the directions Δ_q for $0.9996 \leq q \leq 1.0036$. The resulting grammians all lie in the affine subspace $I - t\Delta + s\Upsilon$. Figure 2.4 shows the resulting grammians in the coordinates (s, t). In the figure, we clearly see the separation between the sets $\mathcal{S}_z^w$ and $\mathcal{D}_z^w$. Figure 2.4 suggests an interesting geometrical relationship between the convex sets $\mathcal{S}_z^w$ and $\mathcal{D}_z^w$. Namely, the sets are extremely close, perhaps identical, over large regions of their boundaries. The sets differ near the approximately hyperextremal element Ξ^*.

TABLE 2.2. 4×4 hyperextremal grammians on the two-holed domain Ω.

N_k	(72, 36, 72)	(96, 48, 96)	(120, 60, 120)	(144, 72, 144)	(168, 84, 168)
$t^*_{N_k} - t^*_{N_{k-1}}$		3.941034×10^{-10}	1.822042×10^{-13}	4.912649×10^{-17}	3.593826×10^{-20}
$s^*_{N_k} - s^*_{N_{k-1}}$		2.406411×10^{-10}	4.112754×10^{-14}	2.830447×10^{-17}	5.915143×10^{-19}
$\Xi_{N_k} - \Xi_{N_{k-1}}$		3.941034×10^{-10}	1.822042×10^{-13}	4.912649×10^{-17}	3.593826×10^{-20}
m_γ^N	2.553699×10^{-11}	$-7.313548 \times 10^{-14}$	$-2.067629 \times 10^{-16}$	$-4.625268 \times 10^{-19}$	$-8.705139 \times 10^{-22}$
$\lambda_{\min}[\gamma(\alpha^*)*\Xi]$	2.553699×10^{-11}	$-1.181303 \times 10^{-14}$	3.185598×10^{-18}	2.324222×10^{-21}	$-4.787487 \times 10^{-24}$
$\lambda_{\min}[\gamma(\beta^*)*\Xi]$	3.185909×10^{-11}	$-7.313548 \times 10^{-14}$	$-2.067629 \times 10^{-16}$	$-4.625268 \times 10^{-19}$	$-8.705139 \times 10^{-22}$
$\|ImX_1\|_2$	0.002275614	0.002275599	0.002275599	0.002275599	0.002275599
$\|ImX_2\|_2$	0.003189101	0.003189079	0.003189079	0.003189079	0.003189079
$\|ImX_1{}^{N_k} - ImX_1{}^{N_{k-1}}\|_2$		3.509010×10^{-8}	7.763171×10^{-11}	1.447898×10^{-13}	2.487826×10^{-16}
$\|ImX_2{}^{N_k} - ImX_2{}^{N_{k-1}}\|_2$		5.425888×10^{-8}	1.098819×10^{-10}	2.014991×10^{-13}	3.655284×10^{-16}
$1 - \tau^*$	$0.0000322810\!0$	$0.0000322694\!8$	$0.0000322694\!5$	$0.0000322694\!5$	$0.0000322694\!5$
$\|X_1 - X_1^{\text{SDP}}\|$	$0.00283160\!3$	$0.00283143\!7$	$0.00283143\!6$	$0.00283143\!6$	$0.00283143\!6$
$\|X_2 - X_2^{\text{SDP}}\|$	$0.00409476\!1$	$0.00409454\!1$	$0.00409454\!0$	$0.00409454\!0$	$0.00409454\!0$

2.7. PLAUSIBILITY ARGUMENTS

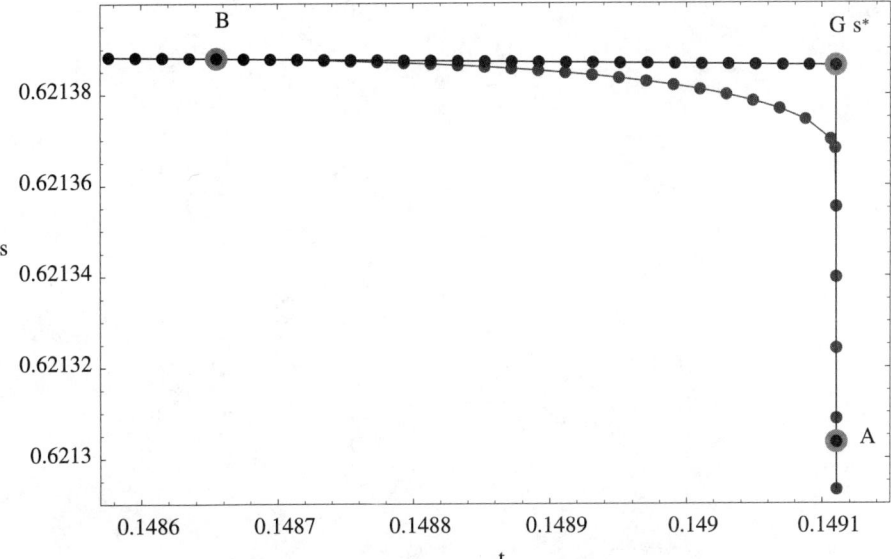

FIGURE 2.4. $\partial \mathcal{S}_z^w$ and $\partial \mathcal{D}_z^w$. The large dot indicates the grammian $G(s^*)$, i.e. the grammian obtained when $q = 1$, point A corresponds to $q = 0.999867$ and point B corresponds to $q = 1.00307$.

CHAPTER 3

Arbitrary Precision Computations of the Poisson Kernel and Herglotz Kernels on Multiply-Connected Circle Domains

3.1. Introduction

In this chapter we shall describe particular aspects of mathematical software that the authors developed for the computation of certain canonical functions associated with a finitely connected domain in the complex plane – namely, the Poisson kernel and certain other functions which we refer to as *Herglotz kernels*. The Poisson kernel is of course famous by virtue of its role in the solution of the Dirichlet problem, and we shall discuss applications of this aspect of our work in a subsequent paper. Herglotz kernels on non-simply connected domains, while firmly established in the theoretical literature, are perhaps less well-known and we now give their definition.

Let Ω be a domain in the complex plane bounded by $m+1$ real analytic Jordan curves $\partial_0, \ldots, \partial_m$. Let T_Ω denote the $m+1$ dimensional torus defined by

$$(3.1.1) \qquad T_\Omega = \partial_0 \times \cdots \times \partial_m.$$

Fix a point $z_0 \in \Omega$. It turns out that for each choice of points $\alpha = (\alpha_0, \ldots, \alpha_m) \in T_\Omega$ there exists a unique $m+1$-valent mapping Γ_α from Ω onto the right half plane $\mathbb{H}$ that has poles at the points $\alpha_0, \ldots, \alpha_m$ and is normalized so that $\Gamma_\alpha(z_0) = 1$. These functions, Γ_α, are referred to as the *Herglotz kernels of the first kind*, so named because of the following theorem. In the theorem, $\mathcal{C}$ denotes the set of holomorphic functions, f, on Ω such that

$$(3.1.2) \qquad \operatorname{Re} f \geq 0$$

and

$$(3.1.3) \qquad f(z_0) = 1$$

THEOREM 3.1.4. *(cf. Chapter 1)* $f \in \mathcal{C}$ *if and only if there exists a probability measure μ supported on T_Ω such that*

$$f(z) = \int_{T_\Omega} \Gamma_\alpha(z) d\mu$$

Note that when $\Omega = \mathbb{D}$ (the open unit disk centered at the origin), and if $z_0 = 0$, then $\Gamma_\alpha(z) = \frac{\alpha_0 + z}{\alpha_0 - z}$, the classical Herglotz kernel, and thus Theorem 3.1.4 in this case is the classical Herglotz Representation Theorem [**Con95**]. This explains why the functions Γ_α are referred to as the Herglotz kernels of the first kind.

However, the functions Γ_α arise in other contexts as well. For example, from the perspective of functional analysis, the functions Γ_α are the extreme points of the set $\mathcal{C}$ defined above ([**Hei85**], [**For79**], Chapter 1). From the point of view of function theory, the functions $\varphi_\alpha = \frac{\Gamma_\alpha - 1}{\Gamma_\alpha + 1}$ are the single-valued Blaschke products of minimal degree that vanish at z_0 [**Fis83**]. From the point of view of classical complex analysis, if $\mathcal{N} = \{f \in Hol(\Omega) : f : \Omega \to \mathbb{H}, f(z_0) = 1\}$ and $\xi \in \Omega$, then the solutions to the extremal problem of the form

$$\sup_{f \in \mathcal{N}} |f'(\xi)| \tag{3.1.5}$$

are Herglotz kernels of the first kind. As such, they comprise a generalization of the classical Ahlfors functions ([**FK99**] and Chapter 1).

It turns out that Theorem 3.1.4 is not the only way to generalize the Herglotz Representation Theorem to the nonsimply connected domain Ω. A different kernel, H_λ, the *Herglotz kernel of the second kind*, is employed to give Theorem 3.1.7 below. The characterization of the Herglotz kernel of the second kind is more subtle than that of the Herglotz kernel of the first kind, and we now describe its construction.

For u a continuous function on $\partial \Omega$, let $u^\wedge$ denote the solution to the Dirichlet problem with boundary data u. Let σ_r denote normalized arclength measure on ∂_r and set $\sigma = \sigma_0 + \cdots + \sigma_m$. Let $P_\lambda(z)$ denote the Poisson kernel with respect to σ, i.e. the unique 2-variable function defined for $z \in \Omega$ and $\lambda \in \partial \Omega$ such that

$$u^\wedge(z) = \int_{\partial \Omega} u(\lambda) P_\lambda(z) d\sigma(\lambda) \tag{3.1.6}$$

whenever u is a continuous function on $\partial \Omega$. Note that we choose $d\sigma$ as our measure on $\partial \Omega$ rather than the more customary choice of arc length measure. If $Rat(\Omega^-)$ denotes the rational functions with poles off Ω^-, then it is well-known that $\text{Re}Rat(\Omega^-)|_{\partial \Omega}$ has codimension m in $L^2(d\sigma)$, the Hilbert space of real-valued, square integrable measurable functions on $\partial \Omega$ with inner product given by $\langle f, g \rangle = \int_{\partial \Omega} fg d\sigma$. Let $\varphi_1, \ldots, \varphi_m$ be an orthonormal basis for the orthogonal complement of $\text{Re}Rat(\Omega^-)$ in $L^2(d\sigma)$.

We are now able to define the *Herglotz kernel of the second kind*. For each fixed $\lambda \in \partial \Omega$, the harmonic function

$$u_\lambda(z) = P_\lambda(z) - \sum_{r=1}^m \varphi_r(\lambda) \varphi_r^\wedge(z)$$

has a unique harmonic conjugate $v_\lambda(z)$ on Ω satisfying $v_\lambda(z_0) = 0$. Let $H_\lambda(z) = u_\lambda(z) + i v_\lambda(z)$. With this kernel the following alternate generalization of the Herglotz representation theorem obtains.

THEOREM 3.1.7. *(cf. Chapter 1) f is a holomorphic function on Ω with $\text{Re} f \geq 0$ if and only if there exists a probability measure μ on $\partial \Omega$ and two scalars $a \geq 0$ and $b \in \mathbb{R}$ such that*
  (1) $\int_{\partial \Omega} \varphi_r d\mu = 0$ for $r = 1, \ldots, m$.
  (2) $f(z) = a \int_{\partial \Omega} H_\lambda(z) d\mu(\lambda) + ib$.

By analogy with our remarks following Theorem 3.1.4, we observe that when $\Omega = \mathbb{D}$ and $z_0 = 0$, $H_\lambda(z) = \frac{\lambda + z}{\lambda - z}$, the classical Herglotz kernel, and thus Theorem 3.1.7 is also a generalization of the classical Herglotz Representation Theorem to the non-simply connected domain Ω. In addition, when Ω is an annulus, an explicit series representation for H_λ is obtainable from an orthonormal basis for $L^2(\Omega)$ [**Agl85**], [**Sar65a**].

3.1. INTRODUCTION

The functions H_λ and similar reproducing kernels have other interesting interpretations and applications from the points of view of functional analysis Chapter 1 and classical complex analysis, as well ([**CW67**] and Chapter 4).

The authors became interested in these Herglotz kernels because of their relationship to an important unsolved problem in operator theory, the so called *rational dilation question*. This question asks whether or not every operator on a complex Hilbert space that has a given compact subset K of the plane as a spectral set (a notion that von Neumann introduced in [**vN51**]) has a rational dilation to the boundary of K (a notion that Sz-Nagy introduced for the case when K is the disc in [**SN53**]). Theoretical work in Chapter 1 showed that if M is an $n \times n$ matrix with eigenvalues in Ω, then M has Ω^- as a spectral set if and only if the $m+1$ parameter system of matrix inequalities

$$(3.1.8) \qquad Re\Gamma_\alpha(M) \geq 0 \text{ for all } \alpha \in T_\Omega$$

obtains.

Likewise, M has a rational dilation to the boundary of Ω if and only if

There exist self-adjoint matrices $\Phi_1, \ldots, \Phi_m$ such that

$$(3.1.9) \qquad ReH_\lambda(M) + \sum_{r=1}^{m} \varphi_r(\lambda)\Phi_r \geq 0 \text{ for all } \lambda \in \partial\Omega.$$

Thus, the *rational dilation question*, which is quite abstract from the point of view of operator theory, can be interpreted quite concretely as the issue of whether every $n \times n$ matrix M which satisfies the matrix inequalities in (3.1.8) gives rise to a solvable parameterized system of linear matrix inequalities as in (3.1.9). The authors decision to approach the issue was prompted by the failure to resolve the rational dilation question theoretically. For example, the only known cases are when $m = 0$ [**SN53**], $m = 1$ [**Agl85**], or when $n = 2$ [**Mis84**], [**Pau87**], [**Agl90**]. Accordingly, most of the authors experimental investigations centered on the cases where $m = 2$ and $n \geq 3$ (Chapter 3 and [**Rap02**]).

Except for the cases when $\Omega = \mathbb{D}$ and when Ω is an annulus, there are no simple formulas for $\varphi_1, \ldots, \varphi_m$, H_λ and Γ_α. Our explorations of the rational dilation question required algorithms for computing these functions. The particular nature of conditions (3.1.8) and (3.1.9), led us to seek algorithms that allow us to compute Γ_α, H_λ, and $\varphi_1, \ldots, \varphi_m$ to high levels of precision at reasonable costs in time and memory. Indeed, we have observed experimentally that the distance in the Hausdorff metric between the set of matrices M that satisfy (3.1.8) and (3.1.9) is extremely small. This observation reveals the infeasibility of using a lengthy string of error estimates to rigorously justify a computational answer to the rational dilation question. Thus, we have focused our efforts on the generation of reliable empirical evidence that has, in turn, allowed us to hypothesize an answer with high confidence (Chapter 3 and [**Rap02**]). This empirical approach naturally places stringent requirements on our algorithms for computing $\varphi_1, \ldots, \varphi_m$, H_λ, and Γ_α. Specifically, the algorithms that we present in this memoir satisfy three main requirements: theoretical error estimates show that, in principle, the functions can be computed to arbitrary accuracy and precision; numerical error can be minimized (or eliminated) by using exact arithmetic during most (or all) of the computations; and experimental results show convergence trends that give us confidence in the accuracy of our algorithms above and beyond the theoretical predictions.

3.2. Computation of the Functions

Based on our requirement that the functions φ_r, P_λ, H_λ, and Γ_α be calculated at arbitrarily high levels of precision, we propose the following algorithm for the computation of these functions: we expand each of the functions in terms of a basis, where for a particular choice of the domain Ω, the coefficients in the basis expansion are computable using exact arithmetic. This allows one to obtain any desired degree of accuracy by including more terms in the basis expansion, and a theoretical estimate of the error in approximation is readily obtainable from the truncation error introduced by including only finitely many terms in the basis expansion.

This section is divided into five subsections. In section 3.2.1, we describe the domains Ω that we consider and our choice of basis on these domains. In section 3.2.2, we use our basis to compute the functions φ_r. In section 3.2.3, we describe the computation of P_λ. Our basis is chosen such that the harmonic conjugate of each basis function is computable exactly. We use this property of our basis in section 3.2.4 to obtain a formula for H_λ. Finally in section 3.2.5, we describe the computation of Γ_α.

3.2.1. A Basis for $L^{2,h}(d\sigma)$. Recall from the introduction that σ denotes normalized arclength measure on the components $\partial_0, \ldots, \partial_m$ of $\partial\Omega$. $L^2(d\sigma)$ is the Hilbert space of real-valued square integrable measurable functions on $\partial\Omega$ with inner product given by $\langle f, g \rangle = \int_{\partial\Omega} fg d\sigma$. Let $L^{2,h}(d\sigma)$ be the closure in $L^2(d\sigma)$ of $ReRat(\Omega^-)|_{\partial\Omega}$. We shall assume that $\partial\Omega = \partial_0 \cup \cdots \cup \partial_m$, where ∂_0 is the unit circle centered at the origin, and $\partial_r \subset \mathbb{D}$ for $r = 1, \ldots, m$. For $r \geq 1$ we let c_r denote the center of ∂_r and let ρ_r denote the radius.

For $r = 0, \ldots, m$ and $l \geq 1$, define rational functions

$$(3.2.1) \qquad F_{r,l}(z) = \begin{cases} z^l, & \text{if } r = 0, \\ \left(\frac{\rho_r}{z - c_r}\right)^l, & \text{if } r \geq 1 \end{cases}$$

and for all $r = 0, \ldots, m$ and $l \geq 1$ define real-valued rational functions

$$(3.2.2) \qquad f_{r,2l-1}(z) = ReF_{r,l}$$
$$(3.2.3) \qquad f_{r,2l}(z) = ImF_{r,l}.$$

Additionally we define the constant function, $f_{0,0}(z) \equiv \frac{1}{\sqrt{2(m+1)}}$ on $\partial\Omega$. Evidently, Runge's Theorem [**Rud87**] implies that the functions $f_{r,k}(z)$ have dense linear span in $L^{2,h}(d\sigma)$.

We enumerate the set of functions $\{f_{r,k} : r = 0, \ldots, m, k \geq 1\}$ by fixing an $m+1$ tuple of positive *even* integers $\mathbf{N} = (N_0, \ldots, N_m)$, letting

$$(3.2.4) \qquad N = N_0 + N_1 + \cdots + N_m,$$

and ordering the functions $f_{r,k}$ into "blocks of size N" as follows. Set $f_1, \ldots, f_N$ equal to

$$(3.2.5) \qquad f_{0,1}, \ldots, f_{0,N_0}, f_{1,1}, \ldots, f_{1,N_1}, \ldots, f_{m,1}, \ldots, f_{m,N_m},$$

respectively, and set $f_{N+1}, \ldots, f_{2N}$ equal to

$$(3.2.6) \qquad f_{0,N_0+1}, \ldots, f_{0,2N_0}, f_{1,N_1+1}, \ldots, f_{1,2N_1}, \ldots, f_{m,N_m+1}, \ldots, f_{m,2N_m},$$

respectively, etc. Let $f_0 = f_{0,0} \equiv \frac{1}{\sqrt{2(m+1)}}$. Let $\mathbf{f}$ denote the infinite column vector

(3.2.7) $$\mathbf{f} = (f_0, f_1, \ldots, f_k, \ldots)^t$$

where t denotes the transpose, and for each positive integer n, let $\mathbf{f_n}$ denote the finite column vector

(3.2.8) $$\mathbf{f_n} = (f_0, f_1, \ldots, f_n)^t.$$

Let $\mathcal{M}_n$ denote the linear span of the functions $f_0, \ldots, f_n$ in $L^{2,h}(d\sigma)$, and let $P_{\mathcal{M}_n}$ denote the orthogonal projection of $L^{2,h}(d\sigma)$ onto $\mathcal{M}_n$.

Given a function $g \in L^{2,h}(d\sigma)$, we wish to evolve a formula for calculating $P_{\mathcal{M}_n} g$ in terms on the functions $f_0, \ldots, f_n$. To achieve this goal, let $\mathbf{u_n} = (u_0, u_1, \ldots, u_n)$ be the vector of functions that results from Gram-Schmidt orthogonalization of the sequence $f_0, f_1, \ldots, f_n$. Let G_n be the $(n+1) \times (n+1)$ grammian matrix $G_n = [\langle f_j, f_i \rangle]_{i,j=0}^n$, and let $G_n^{-1} = C^t C$ be the Cholesky decomposition of G_n^{-1}. Evidently, with these notations, we have that $\mathbf{u_n} = C \mathbf{f_n}$. Hence, if $g \in L^{2,h}(d\sigma)$ then

(3.2.9) $$P_{\mathcal{M}_N} g = \sum_{k=0}^{N} \langle g, u_k \rangle u_k = \sum_{k=1}^{N+1} \langle g, (C\mathbf{f_N})_k \rangle (C\mathbf{f_N})_k = \langle g, \mathbf{f_N} \rangle G_N^{-1} \mathbf{f_N},$$

where $\langle g, \mathbf{f_N} \rangle$ denotes the row vector $(\langle g, f_0 \rangle, \langle g, f_1 \rangle, \ldots, \langle g, f_N \rangle)$.

Additionally, we define the infinite grammian $G = [\langle f_j, f_i \rangle]_{i,j=0}^{\infty}$. We can compute the inner products $\langle f_j, f_i \rangle$, the entries of G (and hence of G_n), exactly using residue theory, as shown in Appendix B.

3.2.2. Computation of the Functions φ_r. Recall that $L^{2,h}(d\sigma)$ has codimension m in $L^2(d\sigma)$ and that the functions $\varphi_1, \ldots, \varphi_m$ form an orthonormal basis for the orthogonal complement of $L^{2,h}(d\sigma)$ in $L^2(d\sigma)$. Our strategy for computing φ_r consists of the following steps.

(1) Pick m functions $g_1, \ldots, g_m \in L^2(d\sigma)$ which are linearly independent modulo $L^{2,h}(d\sigma)$.
(2) Project $g_1, \ldots, g_m$ onto $(L^{2,h}(d\sigma))^\perp$, obtaining m linearly independent functions $\psi_1, \ldots, \psi_m$. Because of Step 1, $\{\psi_1, \ldots, \psi_m\}$ necessarily span $(L^2(d\sigma))^\perp$
(3) Apply the Gram-Schmidt procedure to the set $\{\psi_1, \ldots, \psi_m\}$ obtaining a set of functions $\{\varphi_1, \ldots, \varphi_m\}$.

In Step 1, there are two natural choices for the functions $g_1, \ldots, g_m$. One choice is to set $g_r(z) = 1$ if $z \in \partial_r$ and 0 otherwise (i.e. g_r is harmonic measure for ∂_r). Another choice is to set $g_r = \frac{\log|(z - c_r)|}{\log \rho_r}$. Standard function theory of multiply connected domains implies that both these sets of functions are linearly independent modulo $L^{2,h}(d\sigma)$.

The advantage of choosing the g_r to be harmonic measures is that the inner products $\langle g_r, f_k \rangle$ are computable exactly. The disadvantage is that $g_r^\wedge$, the harmonic extension of g_r to Ω^-, is not computable without numerically solving the Dirichlet problem. However, in our application to the rational dilation question, we require the values of $\varphi_r(z)$ only for $z \in \partial \Omega$. Thus, the choice of harmonic measures for the g_r is optimal.

If one requires the values of $\varphi_r^\wedge$ on the interior of Ω, then the functions

$$\{\frac{\log|(z-c_r)|}{\log \rho_r}\}_{r=1,\ldots,m}$$

are a better choice, as the harmonic extensions are known. The drawback of this choice is that the inner products $\left\langle \frac{\log|(z-c_r)|}{\log \rho_r}, f_k \right\rangle$ cannot be computed using exact arithmetic when $m \geq 2$.

To achieve step 2, we begin with the simple Hilbert space fact that

(3.2.10) $$P_{(L^{2,h}(d\sigma))^\perp} g = g - P_{L^{2,h}(d\sigma)} g.$$

We must approximate $P_{L^{2,h}(d\sigma)} g$ by formula (3.2.9) and a suitable choice of subspace $\mathcal{M}_n$. The subspaces $\mathcal{M}_n$ have the properties $\mathcal{M}_n \subseteq \mathcal{M}_{n+1}$ and $\bigcup_{n=1}^\infty \mathcal{M}_n$ is dense in $L^{2,h}(d\sigma)$. Thus, Hilbert space theory implies that if $g \in L^2(d\sigma)$, $P_{\mathcal{M}_n} g \to P_{L^{2,h}(d\sigma)} g$ in L^2. However, in the present case the convergence of $P_{\mathcal{M}_n} g$ to $P_{L^{2,h}(d\sigma)} g$ is in fact much stronger than mere L^2 convergence. To describe this convergence, let $C^\infty(\partial\Omega)$ denote the infinitely differentiable real-valued functions on $\partial\Omega$ and for $g \in C^\infty(\partial\Omega)$, let $g^{(k)}$ denote the kth derivative of g. If $\{g_n\}$ is a sequence in $C^\infty(\partial\Omega)$ and $g \in C^\infty(\partial\Omega)$, recall that one says that $g_n \to g$ in $C^\infty(\partial\Omega)$ provided

(3.2.11) $$\lim_{n\to\infty} \max_{\lambda \in \partial\Omega} |g^{(k)}(\lambda) - g_n^{(k)}(\lambda)| = 0.$$

Let us agree to say that $g_n \to g$ in $C^\infty(\partial\Omega)$ *exponentially fast* if there exists a positive constant $\eta < 1$ such that for all $k \geq 0$ there exists a constant a_k such that

(3.2.12) $$\max_{\lambda \in \partial\Omega} |g^{(k)}(\lambda) - g_n^{(k)}(\lambda)| \leq a_k \eta^n.$$

Note that $\mathcal{M}_n$ is spanned by functions in $C^\infty(\partial\Omega)$, so tautologically, if $g \in L^2(d\sigma)$ then $P_{\mathcal{M}_n} g \in C^\infty(\partial\Omega)$. It turns out (but is more difficult to see) that it is also true that if g is real-analytic (or C^k or C^∞), then $P_{L^{2,h}(d\sigma)} g$ is real-analytic (resp. C^k or C^∞). Furthermore, if g is real-analytic, then $P_{\mathcal{M}_n} g \to P_{L^{2,h}(d\sigma)} g$ in $C^\infty(\partial\Omega)$ exponentially fast. These facts are proven in Appendix A.

We conclude that for our choices of functions g_r, $P_{L^{2,h}(d\sigma)} g_r$ is well approximated by $P_{\mathcal{M}_n} g_r$ for n sufficiently large. From formula (3.2.9), we obtain the approximate

(3.2.13) $$\psi_r^{\mathbf{N}} = g_r - \langle g_r, \mathbf{f_N} \rangle G_N^{-1} \mathbf{f_N}.$$

By Proposition A.55,

(3.2.14) $$\psi_r^{\mathbf{N}} \to \psi_r \text{ exponentially fast in } C^\infty.$$

Finally, in Step 3 we obtain approximates $\{\varphi_1^{\mathbf{N}}, \ldots, \varphi_m^{\mathbf{N}}\}$ by applying the Gram-Schmidt procedure to the functions $\psi_1^{\mathbf{N}}, \ldots, \psi_m^{\mathbf{N}}$. Since we are applying the Gram-Schmidt procedure to a a finite number of functions whose inner products can be calculated either using exact arithmetic (when the functions g_r are harmonic measures) or to any desired degree of precision (when $g_r = \frac{\log|(z-c_r)|}{\log \rho_r}$) and since the functions $g_1, g_2, \ldots, g_m$ are linearly independent modulo $L^{2,h}(d\sigma)$, the error in calculating $\varphi_r^{\mathbf{N}}$ is bounded by a fixed linear combination (independent of $\mathbf{N}$, for sufficiently large N) of the sup norm of the errors in the functions $\psi_1^{\mathbf{N}}, \ldots, \psi_m^{\mathbf{N}}$. Thus, by virtue of (3.2.14), we have for $r = 1, \ldots, m$

(3.2.15) $$\varphi_r^{\mathbf{N}} \to \varphi_r \text{ exponentially fast in } C^\infty.$$

In particular, for either choice of functions $g_1, \ldots, g_m$, the error in the calculation of the φ_r's is controlled only by N, the number of functions in our basis.

3.2.3. The Computation of P_λ. A well-known formula for P_λ can be calculated from an orthonormal basis $\{v_k\}_{k=1}^\infty$ for $L^2(d\sigma)$ by the formula

$$P_\lambda(z) = \sum_{k=1}^\infty v_k(\lambda) v_k^\wedge(z). \tag{3.2.16}$$

In particular, if $\{u_k\}$ and $\{\varphi_r\}$ are as defined in Section 3.2.1, then $\{\varphi_1, \ldots, \varphi_m\} \cup \{u_k\}_{k=0}^\infty$ is an orthonormal basis for $L^2(d\sigma)$, and hence

$$P_\lambda(z) = \sum_{r=1}^m \varphi_r(\lambda)\varphi_r^\wedge(z) + \sum_{k=0}^\infty u_k(\lambda) u_k^\wedge(z). \tag{3.2.17}$$

Therefore, the function $\tilde{P}_\lambda^n(z)$ defined by

$$\tilde{P}_\lambda^n(z) = \sum_{r=1}^m \varphi_r(\lambda)\varphi_r^\wedge(z) + \sum_{k=0}^n u_k(\lambda) u_k^\wedge(z) \tag{3.2.18}$$

gives an approximation to $P_\lambda(z)$. Furthermore, if z is confined to a compact subset K of Ω, then for each $\lambda \in \partial\Omega$, $\tilde{P}_\lambda^n(z) \to P_\lambda(z)$ exponentially fast in $C^\infty(K)$. However, as $z \to \partial\Omega$, this convergence becomes increasingly sluggish – that is, for a fixed n, (3.2.18) can be very inaccurate if z is sufficiently close to $\partial\Omega$. See Section 3.3.2.1 for details.

To address the issue of calculating $P_\lambda(z)$ when $\lambda \in \partial_r$ and z is close to or in ∂_r, we employ an alternative method of computing P_λ, which uses a "pole-matching" technique. The idea of this approach is that rather than expanding P_λ directly in terms of our basis $\{u_k\}_{k=0}^\infty \cup \{\varphi_r\}_{r=1}^m$, we first subtract a function with a pole at λ, and whose residue at λ equals the residue of P_λ at λ. The function we choose for this purpose is the Poisson kernel for a domain in $\mathbb{C}$ obtained from Ω by filling in the m holes complementary to ∂_r.

Specifically, associate with each boundary component ∂_r, a disk D_r in the extended complex plane as follows. Let D_0 be the unit disk $\mathbb{D}$, and for $r \neq 0$, let D_r be the disk with boundary ∂_r containing the point at ∞. Let $p_\lambda^{D_r}(z)$ denote the Poisson kernel for D_r at λ. Specifically, $p_\lambda^{D_0} = p_\lambda^\mathbb{D} = \text{Re}\left[\frac{\lambda+z}{\lambda-z}\right]$, the Poisson kernel for the unit disk $\mathbb{D}$, and set $P_\lambda^{D_0}(z) = \frac{\lambda+z}{\lambda-z}$. If $r \neq 0$, let $\gamma : D_r \to \mathbb{D}$ be the conformal map given by $\gamma(z) = \frac{\rho_r}{z - c_r}$. Then we can obtain a formula for the function $p_\lambda^{D_r}(z)$ for $r \neq 0$ by taking the composition $p_{\gamma(\lambda)}^\mathbb{D}(\gamma(z))$, giving

$$p_\lambda^{D_r}(z) = \text{Re}\left[\frac{\gamma(\lambda) + \gamma(z)}{\gamma(\lambda) - \gamma(z)}\right] = \text{Re}\left[\frac{\lambda + z - 2c_r}{z - \lambda}\right]. \tag{3.2.19}$$

Define $P_\lambda^{D_r}$, a holomorphic function on Ω whose real part is $p_\lambda^{D_r}$ by

$$P_\lambda^{D_r}(z) = \frac{\lambda + z - 2c_r}{z - \lambda}. \tag{3.2.20}$$

Fix $\lambda \in \partial_r$ and define

$$Q_\lambda^r(z) = \begin{cases} 0, & \text{if } z \in \partial_r \\ p_\lambda^{D_r}(z), & \text{if } z \in \partial_s, s \neq r \end{cases} \tag{3.2.21}$$

Now as a function of z, $(P_\lambda(z) - p_\lambda^{D_r}(z))$, is harmonic on Ω, and the pole of $P_\lambda(z)$ at $z = \lambda$ is exactly canceled by the pole of $p_\lambda^{D_r}(z)$ at $z = \lambda$. Hence, $(P_\lambda(z) - p_\lambda^{D_r}(z))$ extends harmonically to a neighborhood of Ω^-, the closure of Ω. Also,

$$(3.2.22) \qquad (P_\lambda(z) - p_\lambda^{D_r}(z)) = \begin{cases} 0 \text{ for } z \in \partial_r \\ -p_\lambda^{D_r}(z) \text{ for } z \in \partial_s, s \neq r. \end{cases}$$

We conclude that $P_\lambda(z) - p_\lambda^{D_r}(z) = -Q_\lambda^r(z)$ for all $z, \lambda \in \partial\Omega$. Hence

$$(3.2.23) \qquad \left(P_\lambda - \sum_{s=1}^m \varphi_s(\lambda)\varphi_s^\wedge(z)\right) - p_\lambda^{D_r}(z) = -Q_\lambda^r(z) - \sum_{s=1}^m \varphi_s(\lambda)\varphi_s^\wedge(z).$$

Recall from Section 3.1, that

$$\left(P_\lambda - \sum_{s=1}^m \varphi_s(\lambda)\varphi_s^\wedge(z)\right) = ReH_\lambda(z)$$

and that $ReH_\lambda(z)$ is a harmonic function on Ω that possesses a single-valued harmonic conjugate on Ω. This implies that

$$Q_\lambda^r(z) + \sum_{s=1}^m \varphi_s(\lambda)\varphi_s^\wedge(z)$$

is a harmonic function with a single-valued harmonic conjugate on Ω. Hence, for each fixed $\lambda \in \partial\Omega$, if $z \in \partial\Omega$ then
(3.2.24)
$$Q_\lambda^r(z) + \sum_{s=1}^m \varphi_s(\lambda)\varphi_s(z) = P_{L^{2,h}(d\sigma)}\left[Q_\lambda^r(z) + \sum_{s=1}^m \varphi_s(\lambda)\varphi_s(z)\right] = P_{L^{2,h}(d\sigma)}Q_\lambda^r(z),$$

where the last equality follows from the fact that $\varphi_s \in (L^{2,h}(d\sigma))^\perp$.

We obtain, for all $\lambda \in \partial_r$, and all $z \in \Omega^-$

$$(3.2.25) \qquad P_\lambda(z) = p_\lambda^{D_r}(z) + \sum_{s=1}^m \varphi_s(\lambda)\varphi_s^\wedge(z) - \left(P_{L^{2,h}(d\sigma)}Q_\lambda^r\right)^\wedge(z)$$

$$(3.2.26) \qquad \approx p_\lambda^{D_r}(z) + \sum_{s=1}^m \varphi_s^{\mathbf{N}}(\lambda)\left(\varphi_s^{\mathbf{N}}\right)^\wedge(z) - \langle Q_\lambda^r, \mathbf{f_N}\rangle G_N^{-1}\mathbf{f_N}(z),$$

the last approximation resulting from formula (3.2.9). We can compute the inner products $\langle Q_\lambda^r, f_k\rangle$ symbolically using residue theory, obtaining for $k = 0, \ldots, N$, $\langle Q_\lambda^r, f_k\rangle = q_k(\lambda)$, a real-valued function of λ. Let $\mathbf{q_N}(\lambda) = (q_0(\lambda), q_1(\lambda), \ldots, q_N(\lambda))$. Then our formula for an approximation to $P_\lambda(z)$ is
(3.2.27)
$$P_\lambda^{\mathbf{N}}(z) = p_\lambda^{D_r}(z) + \sum_{s=1}^m \varphi_s^{\mathbf{N}}(\lambda)(\varphi_s^{\mathbf{N}})^\wedge(z) - \mathbf{q_N}(\lambda)G_N^{-1}\mathbf{f_N}(z), \text{ for } \lambda \in \partial_r, r = 0, \ldots, m.$$

Note that if $\lambda \in \partial_r$ and $z \in \partial\Omega$,
(3.2.28)
$$P_\lambda(z) - P_\lambda^{\mathbf{N}}(z) =$$
$$\sum_{s=1}^m \varphi_s(\lambda)\varphi_s(z) - \sum_{s=1}^m \varphi_s^{\mathbf{N}}(\lambda)\varphi_s^{\mathbf{N}}(z) + \left(P_{\mathbf{M}_N}Q_\lambda^r\right)(z) - \left(P_{L^{2,h}(d\sigma)}Q_\lambda^r\right)(z).$$

By (3.2.15) and part (1) of Proposition A.82, for each $\lambda \in \partial\Omega$,

(3.2.29) $\quad P_\lambda - P_\lambda^{\mathbf{N}} \to 0$ exponentially fast in $C^\infty(\partial\Omega)$, uniformly in λ,

i.e., with constants a_k and η independent of λ.

3.2.4. The Computation of H_λ. Recall that ReH_λ was defined by the requirement that

(3.2.30) $$ReH_\lambda(z) = P_\lambda(z) - \sum_{r=1}^{m} \varphi_r(\lambda)\varphi_r^\wedge(z),$$

where P_λ denotes the Poisson kernel at λ with respect to the measure σ. Using formula (3.2.27) for $P_\lambda^{\mathbf{N}}$ and the relation (3.2.30), we immediately obtain a series approximation for ReH_λ by

(3.2.31) $\quad ReH_\lambda(z) \approx p_\lambda^{D_r}(z) - \mathbf{q_N}(\lambda)G_N^{-1}\mathbf{f_N}(z)$ for $\lambda \in \partial_r, r = 0, \ldots, m$.

We wish to obtain a series approximation for H_λ by adding to the above formula the harmonic conjugate of its right hand side. At this point, we realize a great advantage of the particular series approximation for the Poisson kernel (3.2.27) that we have utilized. Since ReH_λ is computed in terms of the functions f_k and we are able to calculate a harmonic conjugate of each f_k symbolically, there is no error introduced in passing from ReH_λ to H_λ.

Specifically, according to the definition of our basis given in (3.2.2) and (3.2.3), if f_j is an element of our basis then for all $k \geq 1$, f_{2k} is a harmonic conjugate of f_{2k-1}, and $-f_{2k-1}$ is a harmonic conjugate of f_{2k}. So for each $k \geq 1$, we define a holomorphic function $\hat{F}_k$ on Ω by

$$\hat{F}_0 = \frac{1}{\sqrt{2m+1}}$$
$$\hat{F}_{2k-1} = f_{2k-1} + if_{2k}$$
$$\hat{F}_{2k} = f_{2k} - if_{2k-1}.$$

Thus, for all $j \geq 0$, $f_j = Re\hat{F}_j$. Equivalently, with these notations and by the indexing procedure described in Section 3.2.1, the functions $\hat{F}_k$ are merely the partition into blocks of size N of the functions $\hat{F}_{r,k}$ defined for $r = 0, \ldots, m$ and $l \geq 1$ by

$$\hat{F}_{0,0} = \frac{1}{\sqrt{2m+1}}$$
$$\hat{F}_{r,2l-1} = F_{r,l}$$
$$\hat{F}_{r,2l} = -iF_{r,l}.$$

Let $\widehat{\mathbf{F_N}} = (\hat{F}_0, \ldots, \hat{F}_N)^t$ and $c_{\mathbf{N}}(\lambda) = Im\left[P_\lambda^{D_r}(z_0) - \mathbf{q_N}(\lambda)G_N^{-1}\widehat{\mathbf{F_N}}(z_0)\right]$. We obtain an approximation $H_\lambda^{\mathbf{N}}$ for $H_\lambda(z)$ by

(3.2.32) $$H_\lambda^{\mathbf{N}}(z) = P_\lambda^{D_r}(z) - \mathbf{q_N}(\lambda)G_N^{-1}\widehat{\mathbf{F_N}}(z) - c_{\mathbf{N}}(\lambda)i.$$

Note that if $z \in \partial\Omega$, then
(3.2.33)
$$ReH_\lambda^{\mathbf{N}}(z) - ReH_\lambda(z) = P_\lambda^{\mathbf{N}}(z) - P_\lambda(z) + \sum_{s=1}^{m}\varphi_s(\lambda)\varphi_s(z) - \sum_{s=1}^{m}\varphi_s^{\mathbf{N}}(\lambda)\varphi_s^{\mathbf{N}}(z) \longrightarrow 0$$

exponentially fast in $C^\infty(\partial\Omega)$ uniformly in λ, as described at the end of Section 3.2.3. Let T denote the harmonic conjugation operator described above (see Definition A.56 in Appendix A for a formal development). Since $ImH_\lambda(z_0) = 0$ we have, for all $\lambda \in \partial_r$

$$(3.2.34) \quad ImH_\lambda - ImH_\lambda^N = $$
$$[TP_{\mathcal{M}_N}Q_\lambda^r - TP_{L^{2,h}(d\sigma)}Q_\lambda^r] + [(TP_{L^{2,h}(d\sigma)}Q_\lambda^r)^\wedge(z_0) - (TP_{\mathcal{M}_N}Q_\lambda^r)^\wedge(z_0)]$$

By part(2) of Proposition A.82 the first bracketed term converges to zero exponentially fast in $C^\infty(\partial\Omega)$, uniformly in λ. This implies, together with the maximum principle, that the second bracketed term converges to zero exponentially fast, uniformly in λ. Moreover, since for all $\mathbf{N}$, $H_\lambda^N - H_\lambda$ extends to a neighborhood of Ω^-, these observations and the maximum principle imply that for all $k \geq 0$, there exists $\eta \in (0,1)$ and $a_k \in \mathbb{R}$ such that

$$(3.2.35) \quad \sup_{z \in \Omega^-, \lambda \in \partial\Omega} |\frac{d^k}{dz^k}(H_\lambda^N(z) - H_\lambda(z))| \leq a_k \eta^N.$$

This is a holomorphic version of the notion of exponential convergence in C^∞ introduced in Section 3.2.2 (with implied uniformity in λ).

3.2.5. Computation of Γ_α. Recall that T_Ω denotes the $m+1$ dimensional torus $\partial_0 \times \ldots \partial_m$. If $\alpha = (\alpha_0, \ldots, \alpha_m) \in T_\Omega$, define m vectors $\varphi_1(\alpha), \ldots, \varphi_m(\alpha) \in \mathbb{R}^{m+1}$ by the formula

$$(3.2.36) \quad \varphi_r(\alpha) = (\varphi_r(\alpha_0), \ldots, \varphi_r(\alpha_m)) \text{ for } r = 1, \ldots, m.$$

The following result from Chapter 1 provides the essential ingredient for calculating Γ_α.

THEOREM 3.2.37. *For each $\alpha \in T_\Omega$, there is a unique $(m+1)$-tuple of real numbers $\boldsymbol{w}(\alpha) = (w_0(\alpha), \ldots, w_m(\alpha))$ such that*

$$(3.2.38) \quad \boldsymbol{w}(\alpha) \perp \varphi_r(\alpha) \text{ for } r = 1, \ldots, m \text{ and}$$

$$(3.2.39) \quad \sum_{r=0}^{m} w_r(\alpha) = 1.$$

Furthermore, the entries of $\boldsymbol{w}(\alpha)$ are strictly positive.

To see how this theorem gives rise to an effective procedure for calculating Γ_α, define for each $\alpha \in T_\Omega$ a holomorphic function $\tilde{\Gamma}_\alpha$ on Ω by setting

$$(3.2.40) \quad \tilde{\Gamma}_\alpha(z) = \sum_{r=0}^{m} w_r(\alpha) H_{\alpha_r}(z), z \in \Omega.$$

Using (3.2.38), we see that for $z \in \Omega$

$$(3.2.41) \quad Re\tilde{\Gamma}_\alpha(z) = \sum_{r=0}^{m} w_r(\alpha)\left(ReH_{\alpha_r}(z) - \sum_{s=1}^{m}\varphi_s(\alpha_r)\varphi_s^\wedge(z)\right)$$

$$(3.2.42) \quad = \sum_{r=0}^{m} w_r(\alpha) P_{\alpha_r}(z).$$

Since Theorem 3.2.37 guarantees that $w_r(\alpha) > 0$ for each r, it follows that $\tilde{\Gamma}_\alpha(z)$ is an $m+1$ valent mapping from Ω to the right half plane with poles at $\alpha_0, \ldots, \alpha_m$. Thus $\tilde{\Gamma}_\alpha$ and Γ_α agree up to an affine scaling of the right half plane. Specifically,

$$\Gamma_\alpha(z) = \frac{\tilde{\Gamma}_\alpha(z) - iIm\tilde{\Gamma}_\alpha(z_0)}{Re\tilde{\Gamma}_\alpha(z_0)} \tag{3.2.43}$$

Thus we compute Γ_α using formulas (3.2.40), (3.2.43), and the formulas for φ_r and H_λ derived in Sections 3.2.2 and 3.2.4. We define $\boldsymbol{w}^{\mathbf{N}} = \left(w_0^{\mathbf{N}}(\alpha), \ldots, w_m^{\mathbf{N}}(\alpha)\right)$ according to Theorem 3.2.37 to be the $(m+1)$-tuple of real numbers satisfying (3.2.38) with φ replaced by $\varphi^{\mathbf{N}}$. We compute $\Gamma_\alpha^{\mathbf{N}}$, an approximation for Γ_α by replacing $\tilde{\Gamma}_\alpha$ in formula (3.2.43) by

$$\left(\tilde{\Gamma}_\alpha\right)^{\mathbf{N}} = \sum_{r=0}^{m} w_r^{\mathbf{N}}(\alpha) H_{\alpha_r}^{\mathbf{N}}(z) \text{ , for } z \in \Omega. \tag{3.2.44}$$

The error in $\tilde{\Gamma}_\alpha^{\mathbf{N}}$ can be computed in terms of the error in $\boldsymbol{w}^{\mathbf{N}}$ and $H_\lambda^{\mathbf{N}}$ by

$$\tilde{\Gamma}_\alpha(z) - \tilde{\Gamma}_\alpha^{\mathbf{N}}(z) = \tag{3.2.45}$$

$$\sum_{r=0}^{m} w_r(\alpha) \left(H_{\alpha_r}(z) - H_{\alpha_r}^{\mathbf{N}}(z)\right) + \sum_{r=0}^{m} \left(w_r(\alpha) - w_r^{\mathbf{N}}(\alpha)\right) H_{\alpha_r}^{\mathbf{N}}(z).$$

By (3.2.35), the term $\sum_{r=0}^{m} w_r(\alpha) \left(H_{\alpha_r}(z) - H_{\alpha_r}^{\mathbf{N}}(z)\right)$ converges to zero exponentially fast in $C^\infty(\Omega^-)$, uniformly in λ. Although the term

$$\sum_{r=0}^{m} \left(w_r(\alpha) - w_r^{\mathbf{N}}(\alpha)\right) H_{\alpha_r}^{\mathbf{N}}(z)$$

has poles at $z = \alpha_r$ for $r = 0, 1, \ldots, m$, we can make the following observations about its convergence.

(1) On compact subsets K of Ω,

$$\sum_{r=0}^{m} \left(w_r(\alpha) - w_r^{\mathbf{N}}(\alpha)\right) H_{\alpha_r}^{\mathbf{N}}(z)$$

converges to zero exponentially fast in $C^\infty(K)$.
(2) $Re \sum_{r=0}^{m} \left(w_r(\alpha) - w_r^{\mathbf{N}}(\alpha)\right) H_{\alpha_r}^{\mathbf{N}}(z) \longrightarrow 0$ exponentially fast on $\partial\Omega \setminus \{\alpha_0, \ldots, \alpha_r\}$ since $p_\lambda^{D_r} \equiv 0$ for $\lambda \in D_r$.

Computational experiments revealing the error in $Re\Gamma_\alpha$ on $\partial\Omega$ are discussed in the next section.

3.3. Results

3.3.1. The functions φ_r. For these computational results, we fix Ω to be the two-holed domain such that the outer boundary ∂_0 is the unit circle, ∂_1 is a circle with center $c_1 = 0$ and radius $\rho_1 = \frac{1}{10}$, and ∂_2 is a circle with center $c_2 = \frac{1}{2}$ and radius $\rho_2 = \frac{1}{4}$.

Figure 3.1 shows the graphs of φ_1 and φ_2 on each boundary component. We stated in section 3.2.2 that the φ_r's converge in C^∞ exponentially fast. For the $(m+1)$-tuple $\mathbf{N} = (N_0, \ldots, N_m)$, let $\varphi_r^{\mathbf{N}}$ denote the approximation to φ_r obtain by truncating our basis using $\mathbf{N}$, as described in section 3.2.1. Recall that $1 + \sum_{r=0}^{m} N_r$ is the total number of functions in the expansion. Table 3.1 shows the

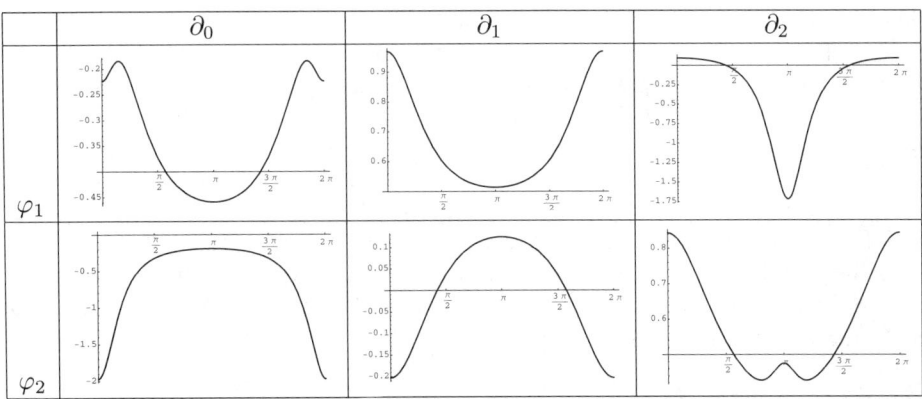

FIGURE 3.1. Graphs of the functions φ_r on boundary components $\partial_0, \partial_1, \partial_2$ with $\mathbf{N} = (76, 36, 68)$, giving a total of $N+1 = 181$ basis functions.

results of an experimental investigation of the behavior of the truncation error $\max_{\lambda \in \partial\Omega} |\varphi_r^{\mathbf{N}}(\lambda) - \varphi_r(\lambda)|, r = 1, 2$. To approximate this error, we have computed $|\varphi_r^{\mathbf{N}}(\lambda) - \varphi_r^{\mathbf{N_0}}(\lambda)|$, where $N_0 \gg N$ and λ ranges through an equally spaced mesh $\mathcal{P}$ consisting of 1000 points on each boundary component ∂_0, ∂_1, and ∂_2. Specifically, we have set $\mathbf{N_0} = (240, 240, 240)$ and computed

(3.3.1) $$E_r^{\mathbf{N}} = \sup_{\lambda \in \mathcal{P}} |\varphi_r^{\mathbf{N}}(\lambda) - \varphi_r^{\mathbf{N_0}}(\lambda)|, r = 1, 2.$$

Of course we assume that $\mathbf{N_0}$ is sufficiently large so that the error

$$\max_{\lambda \in \partial\Omega} |\varphi_r^{\mathbf{N_0}}(\lambda) - \varphi_r(\lambda)|$$

is negligible compared to the values of $E_r^{\mathbf{N}}$ shown in the table. The general trends revealed in Table 3.1 are in good agreement with our theoretical expectation that the errors should converge to zero exponentially fast with $\mathbf{N}$ as discussed in Section 3.2.2. In fact, from the data, we can estimate η, the factor of exponential convergence. We find that $\eta \approx 0.88$. This is a much more favorable rate than predicted by the *a priori* theoretical considerations of Appendix A, which yields an exponential fact at least as large as the constant b defined by formula (A.9). Indeed, for our domain and the 2 : 1 : 2 mix of basis functions in Table 3.1, $b = (.75)^{\frac{1}{2}\frac{2}{5}} \approx 0.944$. For the purposes of our empirical investigation of the rational dilation conjecture, we content ourselves with experimentally derived error estimates, such as those shown in Table 3.1. Although we do not pursue tighter rigorous estimates in this paper, such estimates will presumably be of interest in other applications of these algorithms, particularly when computational efficiency is of high importance. We leave such work for future investigations.

Note that in Table 3.1, for a given value of N only one choice of the $(m+1)$-tuple $\mathbf{N}$ is shown. The question arises: for a fixed valued N, which $(m+1)$-tuple of basis functions, $\mathbf{N} = (N_0, \ldots, N_m)$ with $\sum_{r=0}^{m} N_r = N$, gives the best approximation to φ_r? Unfortunately, the answer to this question is difficult, as the locations of the boundary components relative to one another strongly influence the selection of $\mathbf{N}$. Table 3.2 gives the results of several experiments in this direction. Note that

3.3. RESULTS

TABLE 3.1. Convergence of φ_r

$N+1$	$\mathbf{N}$	$E_1^{\mathbf{N}}$	$E_2^{\mathbf{N}}$
61	$(24, 12, 24)$	0.000608537	0.00123283
121	$(48, 24, 48)$	2.72977×10^{-7}	8.85577×10^{-7}
181	$(72, 36, 72)$	1.29816×10^{-10}	6.3667×10^{-10}
241	$(96, 48, 96)$	6.32681×10^{-14}	4.57661×10^{-13}
301	$(120, 60, 120)$	4.67902×10^{-17}	3.2892×10^{-16}

contrary to what one might expect, for our test domain Ω the best choice is not to choose the entries of $\mathbf{N}$ to be equal. Roughly speaking, there are two sources of error in our basis expansion: truncation error on ∂_r due to the use of finitely many functions of the form $f_{r,n}$ (which become trigonometric functions when restricted to ∂_r), and truncation error due to the finite number of functions of the form $f_{s,n}$ for $s \neq r$ (which decay in magnitude exponentially fast on ∂_r as $n \to \infty$). The first truncation error is directly estimated as error in the "Fourier expansion" on ∂_r. However, the second source of truncation error is intimately connected with the relative locations of the boundary components ∂_r, and is the reason that the optimal choice of $\mathbf{N}$ has unequal entries.

We characterize the qualitative nature of the error in $\varphi_r^{\mathbf{N}}$ as follows. For each $r = 1, \ldots, m$,

$$(3.3.2) \qquad \psi_r^{\mathbf{N}}(\lambda) - \psi_r(\lambda) = (P_{\mathcal{M}_n} g)(\lambda) - (P_{L^{2,h}(d\sigma)} g)(\lambda).$$

In Appendix A, we introduce the notion of an oscillatory sequence $\{g_n\} \subseteq L^2(d\sigma)$. Roughly, this notion means that on each boundary circle $\partial_0, \ldots, \partial_m$ the trigonometric series of $g_n\|_{\partial_r}$ becomes increasingly dominated by high frequency terms as $n \to \infty$ (cf. Definition A.85). By Proposition A.94, $\{\psi_r^{\mathbf{N}}(\lambda) - \psi_r(\lambda)\}$ is oscillatory. Now, be means of Gram-Schmidt orthogonalization, there exists constants $c_{r,s}$ and $c_{r,s}^{\mathbf{N}}$ such that

$$(3.3.3) \qquad \varphi_r = \sum_{s=1}^{m} c_{r,s} \psi_s,$$

and

$$(3.3.4) \qquad \varphi_r^{\mathbf{N}} = \sum_{s=1}^{m} c_{r,s}^{\mathbf{N}} \psi_s^{\mathbf{N}}$$

where $c_{r,s}^{\mathbf{N}} \to c_{r,s}$ as $|\mathbf{N}| \to \infty$. Thus,

$$(3.3.5) \qquad \varphi_r^{\mathbf{N}}(\lambda) - \varphi_r(\lambda) = \sum_{s=1}^{m} \left((c_{r,s}^{\mathbf{N}} - c_{r,s}) \psi_s(\lambda) + (\psi_s^{\mathbf{N}}(\lambda) - \psi_s(\lambda)) c_{r,s}^{\mathbf{N}} \right)$$

$$(3.3.6) \qquad = \sum_{s=1}^{m} c_{r,s}^{\mathbf{N}} \left(\psi_s^{\mathbf{N}}(\lambda) - \psi_s(\lambda) \right) + \sum_{s=1}^{m} (c_{r,s}^{\mathbf{N}} - c_{r,s}) \psi_s(\lambda).$$

Therefore, we see that the error in $\varphi_r^{\mathbf{N}}$ is the sum of two terms, the first of which is oscillatory, the second an element of $L^2(d\sigma) \ominus L^{2,h}(d\sigma)$ of small magnitude.

122 3. ARBITRARY PRECISION COMPUTATIONS

TABLE 3.2. Choosing the proper mix **N** of basis functions.

$N+1$	**N**	$E_1^{\mathbf{N}}$	$E_2^{\mathbf{N}}$
241	$(80, 40, 120)$	6.56216×10^{-12}	5.70341×10^{-11}
241	$(80, 80, 80)$	1.01997×10^{-11}	5.70341×10^{-11}
241	$(96, 48, 96)$	6.32681×10^{-14}	4.57661×10^{-13}
241	$(100, 44, 96)$	2.10903×10^{-13}	1.36974×10^{-13}
241	$(110, 70, 60)$	5.92492×10^{-9}	2.27502×10^{-9}
241	$(140, 60, 40)$	3.53479×10^{-6}	1.2282×10^{-6}

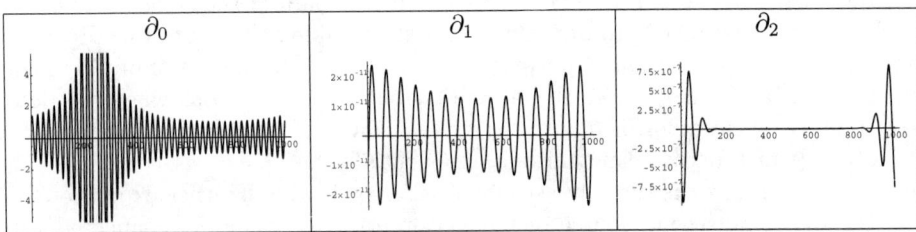

FIGURE 3.2. P_λ^N computed without pole-matching, with $\mathbf{N} = (76, 36, 68)$, $\lambda = e^{\frac{2\pi i}{3}}$.

These two sources of error will become evident when we consider computational experiments with P_λ and Γ_α.

3.3.2. The Function P_λ.

3.3.2.1. *No Pole Matching.* Figure 3.2 shows the graph of $P_\lambda^N(z)$ for $\lambda = e^{\frac{2\pi i}{3}}$, computed using formula (3.2.18) with $\mathbf{N} = (76, 36, 68)$. While the approximation $P_\lambda^N(z)$ is accurate for $z \in \partial_r, r \neq 0$, the error for $z \in \partial_0$ is large. In fact, one can show that the series approximation (3.2.18) for $P_\lambda^N(z)$ does not converge for $z \in \partial_0$.

3.3.2.2. *With Pole Matching.* Figure 3.3 shows the graph of $P_\lambda^N(z)$ for $\lambda = e^{\frac{2\pi i}{3}}$, and $z \in \partial\Omega$ computed with the pole-matching formula (3.2.27) and $\mathbf{N} = (76, 36, 68)$. In contrast to the formula without pole matching, here the error on the boundary components ∂_r is uniformly small.

The qualitative nature of Figure 3.3 can be understood as follows. Note that if $\lambda \in \partial_r$, $z \in \partial\Omega$, and $z \neq \lambda$, then since $P_\lambda(z) = 0$ we have

$$P_\lambda^{\mathbf{N}}(z) = P_\lambda^{\mathbf{N}}(z) - P_\lambda(z)$$
$$= \sum_{r=1}^{m} \varphi_r^{\mathbf{N}}(\lambda)\varphi_r^{\mathbf{N}}(z) - \sum_{r=1}^{m} \varphi_r(\lambda)\varphi_r(z) - (P_{\mathcal{M}_N} Q_\lambda^r)(z) + (P_{L^{2,h}(d\sigma)} Q_\lambda^r)(z)$$
$$= \sum_{r=1}^{m} \left(\varphi_r^{\mathbf{N}}(\lambda) - \varphi_r(\lambda)\right)\varphi_r(z) + \sum_{r=1}^{m} \left(\varphi_r^{\mathbf{N}}(z) - \varphi_r(z)\right)\varphi_r^{\mathbf{N}}(\lambda)$$
$$- \left[(P_{\mathcal{M}_N} Q_\lambda^r)(z) - (P_{L^{2,h}(d\sigma)} Q_\lambda^r)(z)\right].$$

For each fixed $\lambda \in \partial\Omega$, the first summation above is an element of $L^2(d\sigma) \ominus L^{2,h}(d\sigma)$ of small magnitude. The term in brackets is oscillatory by Proposition A.94. The

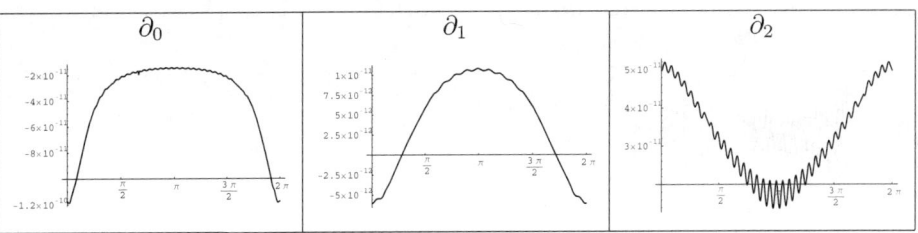

FIGURE 3.3. Poisson kernel $P_\lambda(z)$, computed using pole matching formula (3.2.27) with $\mathbf{N} = (76, 36, 68)$.

second summation is, by the discussion at the end of Section 3.3.1, a linear combination of terms that have an oscillatory component and a small component that lies in $L^2(d\sigma) \ominus L^{2,h}(d\sigma)$. Therefore, for $z \in \partial\Omega$, $z \neq \lambda$, $P_\lambda^\mathbf{N}(z)$ is also the sum of an oscillatory component and a small constant times a function in $L^2(d\sigma) \ominus L^{2,h}(d\sigma)$. This qualitative behavior of the error is clearly seen by comparing Figures 3.1 and 3.3.

3.3.3. The Function H_λ. Table 3.3 shows the results of an experimental investigation of the truncation error $\max_{z \in \partial\Omega} |H_\lambda^\mathbf{N}(z) - H_\lambda(z)|$ for three values of λ. As in Section 3.3.1, we approximate the error by the quantity

$$(3.3.7) \qquad E_\lambda^\mathbf{N} = \max_{z \in \mathcal{P}} |H_\lambda^\mathbf{N}(z) - H_\lambda^{\mathbf{N}_0}(z)|,$$

where $\mathcal{P}$ is a mesh consisting of 1000 points on each boundary component ∂_r and $|\mathbf{N}_0| \gg |\mathbf{N}|$. Again, consistent with our theoretical expectations, we see an exponential decay in the error as $\mathbf{N}$ increases, and the exponential factor η, can be estimated from Table 3.3. We find $\eta \approx 0.88$. Further empirical evidence of the accuracy of $H_\lambda^\mathbf{N}$ can be inferred from the experimental behavior of $\Gamma_\alpha^\mathbf{N}$ described in the next section.

TABLE 3.3. Convergence of H_λ.

$N+1$	$\mathbf{N}$	$E_\lambda^\mathbf{N}$
61	(24, 12, 24)	0.0003710540
121	(48, 24, 48)	2.461572×10^{-7}
181	(72, 36, 72)	1.818228×10^{-10}
241	(96, 48, 96)	1.347517×10^{-13}
301	(120, 60, 120)	9.836160×10^{-17}

3.3.4. The Function Γ_α. For $\alpha = (-i, -\frac{i}{10}, \frac{1}{4})$, $\mathbf{N} = (76, 36, 68)$, and $z_0 = \frac{1}{2} - \frac{i}{2}$, the graphs of $Re\Gamma_\alpha(z)$ and $Im\Gamma_\alpha(z)$ on the boundary components $\partial_0, \ldots, \partial_m$ are shown in Figures 3.4 and 3.5. By definition, the real part of $\Gamma_\alpha(z)$ is zero on $\partial\Omega \setminus \{\alpha_0, \alpha_1, \alpha_2\}$, and the imaginary part has poles at $z = \alpha_r$ and is one-to-one on each ∂_r. The computed Γ_α's exhibit both of these behaviors.

Using formula (3.2.45), the error in $Re\Gamma_\alpha^\mathbf{N}$ can be seen, as in the case of the error in $P_\lambda^\mathbf{N}$ to have an oscillatory component and a component which lies in $L^2(d\sigma) \ominus L^{2,h}(d\sigma)$. This behavior is evident in Figure 3.4. Note that the relative significance of the two types of error varies according to the boundary component ∂_r.

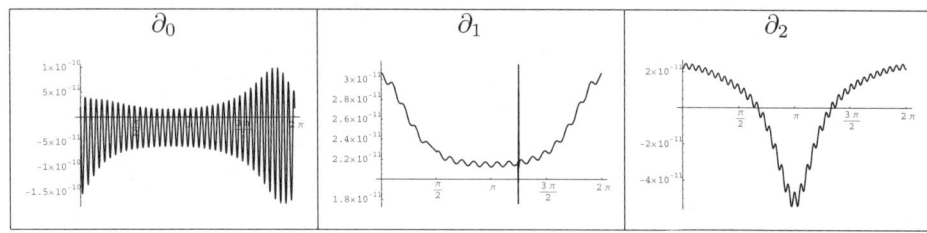

FIGURE 3.4. $Re\Gamma_\alpha(z)$ on $\partial\Omega$

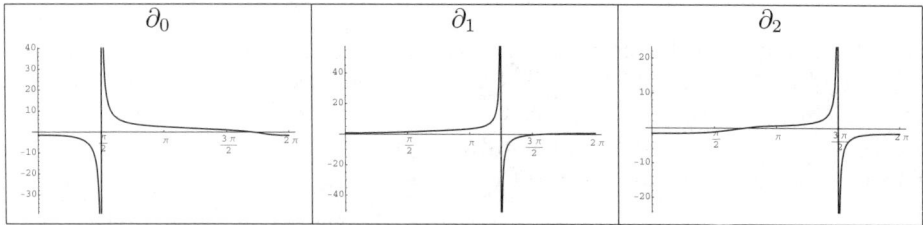

FIGURE 3.5. $Im\Gamma_\alpha(z)$ on $\partial\Omega$

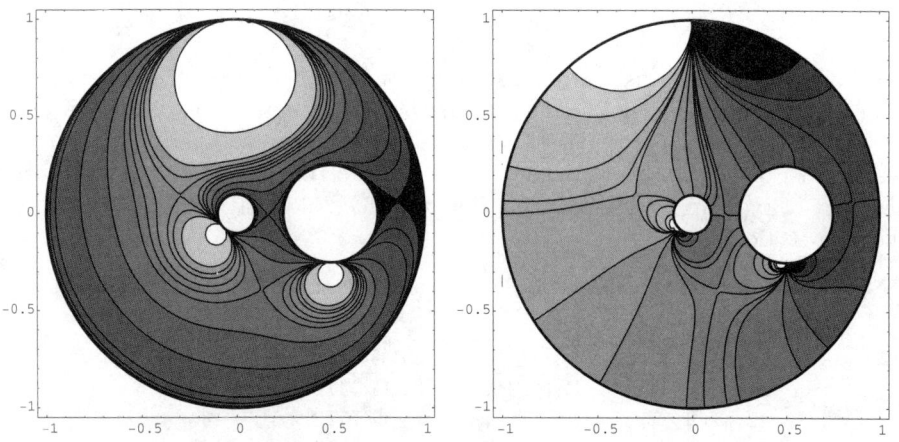

FIGURE 3.6. Contours of $Re\varphi_\alpha(z)$ (left) and $Im\varphi_\alpha(z)$ (right).

In Section 3.1, we noted that $\varphi_\alpha(z) = \frac{\Gamma_\alpha(z)-1}{\Gamma_\alpha(z)+1}$ is a single-valued Blaschke product of minimal degree, and that every Ahlfors function on Ω is given by a φ_α, for some α in T_Ω. Figure 3.3.4 gives contour plots of $Re\varphi_\alpha$, and $Im\varphi_\alpha$ for $\alpha = (i, \frac{1}{10}e^{i\frac{5\pi}{4}}, \frac{1}{2} - \frac{i}{4})$. These contours intersect at the four points $w_1, w_2, w_3, w_4 \in \Omega$ defined by $\varphi'_\alpha = 0$. Methods involving integral equations have been employed to compute Ahlfors functions. In particular, Tegtmeyer [**Teg98**] computes the Ahlfors function by numerically solving the integral equation of Kerzman and Stein [**KS78**]. Our methods for computing φ_α compare favorably with those of Tegtmeyer, but our method has the advantage of yielding a formula from which one can obtain any Ahlfors function by evaluating φ_α with the appropriate choice of α.

3.3.5. A Sample Calculation Related to the Rational Dilation Question.
Finally, we present a sample calculation with the functions Γ_α which relates to the rational dilation question. Recall from (3.1.8) that an $n \times n$ matrix M with eigenvalues contained in Ω has Ω^- as a spectral set, provided $Re\Gamma_\alpha(M) \geq 0$ for all $\alpha \in T_\Omega$.

Particular $n \times n$ matrices, obtained by compressing certain linear operators – referred to in the literature as *bundle shifts* [**AD76**] – to n dimensional invariant subspaces are known to lie on the topological boundary of the set of $n \times n$ matrices which have Ω^- as a spectral set. In terms of condition (3.1.8), this implies that if M is such a compression of a bundle shift, then $Re\Gamma_\alpha(M) \geq 0$ for all $\alpha \in T_\Omega$ and has a kernel for some $\alpha^* \in T_\Omega$. In fact, more is true: the matrix $Re\Gamma_\alpha(M)$ will have a kernel for all $\alpha \in T_\Omega$ (Chapter 1). We examine a 3×3 bundle shift M. We compute the function $(\lambda_{\min}(Re\Gamma_\alpha(M)))$ for $\alpha \in \mathcal{P} \subseteq T_\Omega$ a mesh of points on the $m+1$-torus, computing Γ_α with $\mathbf{N} = (76, 36, 68)$. We find that $\max_{\alpha \in \mathcal{P}}(\lambda_{\min}(Re\Gamma_\alpha(M))) = 1.3474816 \times 10^{-10}$. Similarly when M is taken to be an 8×8 compression of a bundle shift, $\max_{\alpha \in \mathcal{P}}(\lambda_{\min}(Re\Gamma_\alpha(M))) = 3.83621 \times 10^{-10}$. With $\mathbf{N} = (120, 60, 120)$, $\max_{\alpha \in \mathcal{P}}(\lambda_{\min}(Re\Gamma_\alpha(M))) = 4.281718 \times 10^{-16}$. This computation provides further independent evidence of the accuracy of our algorithm for computing Γ_α.

CHAPTER 4

Schwartz Kernels on Multiply Connected Domains

In the literature many examples of representation theorems that are similar to our Theorem 1.5.14 have appeared ([**CW67**], [**Fis83**], [**FK99**], [**Kha84**], [**Zmo58**]). In this chapter we shall discuss a generalization of H_λ, the Herglotz kernel of the second type (Definition 1.4.2), and a corresponding representation of holomorphic functions which encompasses both Theorem 1.4.5 and these alternative representation theorems from the literature. For historical consistency it is natural to refer to this generalized kernel as the *Schwartz kernel*. In Theorem 4.5 below, we derive a formula for any such kernel in terms of the Herglotz kernel of the second type. In addition, in Theorem 4.15 below we provide a general constructive method for evolving the real part of a Schwartz kernel, thereby unifying the various approaches to specific cases of Schwartz kernels that have appeared in the literature. A variant of this construction, which applies to Schwartz kernels whose real parts are sufficiently regular to be induced by measures on $\partial\Omega$, is given in Theorem 4.33. The results and methods of this section are based on the ideas in [**CW67**], [**For79**], [**Kha84**].

We shall begin by observing that the class of functions which can be represented by means of the integral formula in Theorem 1.4.5 can easily be enlarged. Specifically, we define

$$\text{Hol}_{\text{meas}}(\Omega) = \{f \in \text{Hol}(\Omega) : \text{Re} f = \mu^\wedge \text{ for some } \mu \in M_{\mathbb{R}}^h(\partial\Omega)\}.$$

If $f \in \text{Hol}_{\text{meas}}(\Omega)$ then by (1.6.8) there is a unique $\mu \in M_{\mathbb{R}}(\partial\Omega)$ such that $\text{Re} f = \mu^\wedge$; in this case we say that μ is the measure on $\partial\Omega$ induced by $\text{Re} f$. Note that we are not assuming any normalization on the imaginary parts of functions in $\text{Hol}_{\text{meas}}(\Omega)$, as in the case of Theorem 1.4.5, which, as stated, applied only to functions in the set C, defined in Definition 1.3.22.

Because every real Borel measure on $\partial\Omega$ is the difference of two finite positive measures, $\text{Hol}_{\text{meas}}(\Omega)$ is the smallest real subspace of $\text{Hol}(\Omega)$ containing C and the constant functions. Moreover, the representation of functions in C given in Theorem 1.4.5 extends to functions in $\text{Hol}_{\text{meas}}(\Omega)$ in the following way: for every $f \in \text{Hol}_{\text{meas}}(\Omega)$ there exists a constant $b \in \mathbb{R}$ such that

$$(4.1) \qquad f(z) = \int_{\partial\Omega} H_\lambda(z) d\mu(\lambda) + ib$$

where μ is the measure on $\partial\Omega$ induced by $\text{Re} f$.

Recall that our definition of $H_\lambda(z)$ depends on the choice of normalizing function τ in Definition 1.4.2. For simplicity and we shall hereafter in this section fix $z_0 \in \Omega$ and define τ by $\tau(f) = \text{Im} f(z_0)$, so that our normalization on H_λ will be

$$(4.2) \qquad \text{Im} H_\lambda(z_0) = 0 \quad \text{for all } \lambda \in \partial\Omega.$$

Writing $H_\lambda(z)$ in the more conventional kernel notation

$$H_\lambda(z) = k(z,\lambda)$$

(4.1) and (4.2) yield, for every $f \in \text{Hol}_{\text{meas}}(\Omega)$,

(4.3) $$f(z) = \int_{\partial\Omega} k(z,\zeta)d\mu(\zeta) + i\text{Im}\,f(z_0)$$

In the literature representations of the type in (4.3) have been developed and applied to various problems. For example in [**Kha84**] Khavinson uses a formula of this type to construct functions in the Nevanlinna class on Ω and in $H^p(\Omega)$ that have prescribed modulus on $\partial\Omega$. As another example, in [**CW67**] Coifman and Weiss use a similar formula to develop a factorization theory for the Nevanlinna class on Ω as well an expression for a conformal map from Ω onto an annular slit region. The kernels $k(z,\lambda)$ constructed by these authors will be be described after Theorem 4.36 below.

Generally authors have referred to representations of the type in (4.3) as Schwartz integrals (or Schwartz-Stieljes integrals), and the kernels be appearing in such representations as Schwartz kernels. The following definition formalizes the notion of such a kernel.

DEFINITION 4.4. Let k be a complex-valued function defined on $\Omega \times \partial\Omega$. k is a *Schwartz kernel on Ω* if the following conditions hold

(i) For all $\lambda \in \partial\Omega$ $\quad z \mapsto k(z,\lambda)$ is holomorphic on Ω and $\text{Im}\,k(z_0,\lambda) = 0$
(ii) For each $z \in \Omega$ $\quad \lambda \mapsto k(z,\lambda)$ is continuous on $\partial\Omega$
(iii) For all $f \in \text{Hol}_{\text{meas}}(\Omega)$ the reproducing property (4.3) holds, where μ is the measure on $\partial\Omega$ induced by $\text{Re}\,f$.

The following theorem gives a description of the general Schwartz kernel in terms of H_λ, the Herglotz kernel of the second type. Recall that σ is a Borel measure on $\partial\Omega$ which is mutually absolutely continuous with respect to arc length measure. (Generally authors have taken σ to be arc length measure.) Also, recall that the continuous functions $\phi_1,\ldots,\phi_m$ form an orthonormal basis for $L^2_{\mathbb{R}}(\sigma) \ominus L^{2,h}_{\mathbb{R}}(\sigma)$.

THEOREM 4.5. $k : \Omega \times \partial\Omega \to \mathbb{C}$ *is a Schwartz kernel on Ω if and only if there exist $f_1,\ldots,f_m \in \text{Hol}(\Omega)$ such that $0 = \text{Im}\,f_1(z_0) = \cdots = \text{Im}\,f_m(z_0)$ and for all $(z,\lambda) \in \Omega \times \partial\Omega$*

(4.6) $$k(z,\lambda) = H_\lambda(z) + \sum_{r=1}^{m} \phi_r(\lambda) f_r(z)$$

PROOF. Using (1.2.9), (4.1) and (4.2) is easily verified that if $f_1,\ldots,f_m \in \text{Hol}(\Omega)$ satisfy $0 = \text{Im}\,f_1(z_0) = \cdots = \text{Im}\,f_m(z_0)$ and k is given by (4.6), then k is a Schwartz kernel.

Conversely, assume that k is a Schwartz kernel. Recall that for each fixed $\lambda \in \partial\Omega$, $H_\lambda(z) \in \text{Hol}_{\text{meas}}(\Omega)$ and furthermore, the measure on $\partial\Omega$ induced by $\text{Re}\,H_\lambda(z)$ is $\delta_\lambda - \sum_{r=1}^{m} \phi_r(\lambda)\phi_r d\sigma$. By the reproducing property (4.3) and our assumed normalization (4.2), for all λ we have

$$H_\lambda(z) = \int_{\partial\Omega} k(z,\zeta) d\left(\delta_\lambda - \sum_{r=1}^m \phi_r(\lambda)\phi_r\sigma\right)(\zeta)$$
$$= k(z,\lambda) - \sum_{r=1}^m \phi_r(\lambda) \int_{\partial\Omega} k(z,\zeta)\phi_r(\zeta) d\sigma(\zeta).$$

For $r = 1,\ldots,m$ define

(4.7) $$f_r(z) = \int_{\partial\Omega} k(z,\zeta)\phi_r(\zeta) d\sigma(\zeta).$$

Since $\zeta \mapsto k(z,\zeta)$ is continuous for each fixed z and since $\operatorname{Im} k(z_0,\zeta) = 0$ for all $\zeta \in \partial\Omega$, we see that each f_r is a well-defined complex-valued function on Ω satisfying $\operatorname{Im} f_r(z_0) = 0$. Moreover, for all $(z,\lambda) \in \Omega \times \partial\Omega$ (4.6) holds. It remains to be shown that $f_1,\ldots,f_m$ are holomorphic.

Since Definition 4.4 makes no stipulation about the behavior of k with respect to joint variation in z and λ, (4.7) does not immediately imply that each f_r is holomorphic. However, we have equation (4.6) at our disposal. Fix points $\lambda_r \in \partial_r$, $r = 1,\ldots,m$. (4.6) implies that for all $z \in \Omega$,

(4.8) $$\begin{pmatrix} k(z,\lambda_1) - H_{\lambda_1}(z) \\ \vdots \\ k(z,\lambda_m) - H_{\lambda_m}(z) \end{pmatrix} = [\phi_s(\lambda_r)]_{r,s=1}^m \begin{pmatrix} f_1(z) \\ \vdots \\ f_m(z) \end{pmatrix}.$$

By Lemma 1.3.20 the matrix $[\phi_s(\lambda_r)]_{r,s=1}^m$ is nonsingular. Multiplying each side of (4.8) by $([\phi_s(\lambda_r)]_{r,s=1}^m)^{-1}$, we obtain expressions for $f_1(z),\ldots,f_m(z)$ which are linear combinations of $k(z,\lambda_s) - H_{\lambda_s}(z)$, $s = 1,\ldots,m$, the latter of which are holomorphic in z. Hence $f_1,\ldots,f_m$ are holomorphic. This completes the proof of Theorem 4.5. $\square$

Before continuing we remark that Theorem 4.5 implies that every Schwartz kernel is in fact jointly continuous on $\Omega \times \partial\Omega$.

According to Theorem 4.5 every Schwartz kernel can be realized as a straightforward modification of $H_\lambda(z)$. As an immediate consequence we obtain the following result, which points out that $H_\lambda(z)$ enjoys no special status among the set of Schwartz kernels on Ω.

COROLLARY 4.9. If k_1 and k_2 are Schwartz kernels on Ω then there exists $f_1,\ldots,f_m \in \operatorname{Hol}(\Omega)$ such that $0 = \operatorname{Im} f(z_0) = \cdots = \operatorname{Im} f(z_m)$ and

$$k_2(z,\lambda) = k_1(z,\lambda) + \sum_{r=1}^m \phi_r(\lambda) f_r(z).$$

The remainder of this section is devoted to generalizing two classical, more constructive approaches to evolving Schwartz kernels. The first such construction, which we shall call the Coifman–Weiss construction, is the more general of the two and results in all Schwartz kernels on Ω (Theorem 4.15 below). The second, which we shall refer to as the Khavinson construction, results in those Schwartz kernels whose real parts are sufficiently regular to induce measures on $\partial\Omega$ (Theorem 4.33 below). In contrast to Theorem 4.5, which presents the general Schwartz kernel as a modification of the Herglotz kernel of the second type, both constructions modify the Poisson kernel to obtain the real part of a Schwartz kernel.

The Coifman–Weiss construction depends on a choice of m harmonic functions $u_1, \ldots, u_m$ and a homology basis $\Gamma = \{\gamma_1, \ldots, \gamma_m\}$ for Ω. For simplicity we will assume that $\gamma_1, \ldots \gamma_m$ are simple closed analytic curves. Let $\frac{\partial}{\partial n}$ denote the outward normal derivative and ℓ denote arc length measure along $\gamma_1, \ldots, \gamma_m$.

Setting
$$u = (u_1, \ldots, u_m)$$
define the matrix $M_1(u, \Gamma)$ by

(4.10) $$M_1(u, \Gamma) = \left[\int_{\gamma_r} \frac{\partial u_s}{\partial n} d\ell \right]_{r,s=1}^m$$

and functions $\psi_1, \ldots, \psi_m : \partial\Omega \to \mathbb{R}$ by

(4.11) $$\psi_r(\lambda) = \int_{\gamma_r} \frac{\partial P_z^\sigma(\lambda)}{\partial n} d\ell(z).$$

Note that when $\gamma_1, \ldots, \gamma_m$ are homologous in Ω^- to $\partial_1, \ldots, \partial_m$ respectively, and when $\sigma = $ arc length measure on $\partial\Omega$, $\gamma_r(\lambda) = \frac{\partial \omega_r}{\partial n}(\lambda)$, where ω_r is the "harmonic measure" for $\partial_r, r = 1, \ldots, m$. The idea of the construction will be to modify the Poisson kernel, $P_z^\sigma(\lambda)$, at each $\lambda \in \partial\Omega$ by a linear combination of $u_1, \ldots, u_m$ so that the result has a single-valued harmonic conjugate. Explicitly we wish to find scalars $\Lambda_1(\lambda), \ldots, \Lambda_m(\lambda)$ such that

(4.12) $$z \mapsto P_z^\sigma(\lambda) - \sum_{s=1}^m \Lambda_s(\lambda) u_s(z)$$

has a single-valued harmonic conjugate for each $\lambda \in \partial\Omega$. Using the fact that a harmonic function U has a single-valued conjugate if and only if each of the periods $\int_{\gamma_r} \frac{\partial U}{\partial n} d\ell, r = 1, \ldots, m$ vanishes, we shall derive an explicit formula for $\Lambda_1(\lambda), \ldots, \Lambda_m(\lambda)$ in terms of $M_1(u, \Gamma)$ and $\psi_1(\lambda), \ldots, \psi_m(\lambda)$. Indeed, (4.12) has a single-valued conjugate if and only if

$$0 = \int_{\gamma_r} \frac{\partial}{\partial n}(P_z^\sigma(\lambda) - \sum_{s=1}^m \Lambda_s(\lambda) u_s(z)) d\ell(z)$$
$$= \int_{\gamma_r} \frac{\partial P_z^\sigma(\lambda)}{\partial n} d\ell(z) - \sum_{s=1}^m \Lambda_s(\lambda) \int_{\gamma_r} \frac{\partial u_s}{\partial n} d\ell \text{ for all } r = 1, \ldots, m,$$

which is equivalent to the assertion

(4.13) $$\begin{pmatrix} \psi_1(\lambda) \\ \vdots \\ \psi_m(\lambda) \end{pmatrix} = M_1(u, \Gamma) \begin{pmatrix} \Lambda_1(\lambda) \\ \vdots \\ \Lambda_m(\lambda) \end{pmatrix}.$$

Thus, if $M_1(u, \Gamma)$ is nonsingular (4.13) allows us to solve for $\Lambda_1(\lambda), \ldots, \Lambda_m(\lambda)$ explicitly:

(4.14) $$\begin{pmatrix} \Lambda_1(\lambda) \\ \vdots \\ \Lambda_m(\lambda) \end{pmatrix} = M_1(u, \Gamma)^{-1} \begin{pmatrix} \psi_1(\lambda) \\ \vdots \\ \psi_m(\lambda) \end{pmatrix}.$$

The following theorem shows that if $\Lambda_1,\ldots,\Lambda_m$ are defined by (4.14) then (4.12) not only has a single-valued harmonic conjugate for each $\lambda \in \partial\Omega$, but also defines the real part of a Schwartz kernel. Moreover, the theorem establishes that the real part of every Schwartz kernel arises in this manner.

THEOREM 4.15. *(The Coifman–Weiss Construction)*
If $M_1(u,\Gamma)$ is nonsingular and $\Lambda_1,\ldots,\Lambda_m : \partial\Omega \to \mathbb{R}$ are defined by (4.14) then

$$(4.16) \qquad P_z^\sigma(\lambda) - \sum_{s=1}^m \Lambda_s(\lambda) u_s(z)$$

is the real part of a Schwartz kernel on Ω. Conversely, if k is a Schwartz kernel on Ω, then there exists an m-tuple of harmonic functions $u = (u_1,\ldots,u_m)$ such that $M_1(u,\Gamma)$ is nonsingular and $\operatorname{Re} k(z,\lambda)$ is given by (4.16), where $\Lambda_1,\ldots,\Lambda_m$ satisfy (4.14).

The proof of Theorem 4.15 will be based on four preliminary results. Recall that if V is a vector space over a field F and W is a subspace of V then $v_1,\ldots,v_m \in V$ are said to be linearly independent modulo W if the cosets $v_1 + W,\ldots,v_m + W$ are linearly independent in the quotient space V/W (equivalently, if for every nonzero $a \in F^m$, $\sum_{s=1}^m a_s v_s \notin W$).

LEMMA 4.17. *Let $u = (u_1,\ldots,u_m)$ where $u_1,\ldots,u_m \in \operatorname{Harm}_\mathbb{R}(\Omega)$. $u_1,\ldots,u_m$ are linearly independent modulo $\operatorname{Harm}_\mathbb{R}^h(\Omega)$ if and only if $M_1(u,\Gamma)$ is nonsingular.*

PROOF. The lemma follows by noting that if $a \in \mathbb{R}^m$ then

$$\int_{\gamma_r} \frac{\partial}{\partial n} \left(\sum_{s=1}^m a_s u_s \right) d\ell = (M_1(u,\Gamma) a)_r$$

and recalling that a harmonic function v is an element of $\operatorname{Harm}_\mathbb{R}(\Omega)$ if and only if all of the periods $\int_{\gamma_r} \frac{\partial v}{\partial n} d\ell, r = 1,\ldots,m$ vanish. $\square$

In the sequel we shall adopt the notation

$$\phi^\wedge = ((\phi_1 \sigma)^\wedge, \ldots, (\phi_m \sigma)^\wedge).$$

LEMMA 4.18. *$M_1(\phi^\wedge, \Gamma)$ is nonsingular.*

PROOF. By Lemma 4.17 it suffices to show, that $(\phi_1\sigma)^\wedge, \ldots, (\phi_m\sigma)^\wedge$ are linearly independent modulo $\operatorname{Harm}_\mathbb{R}^h(\Omega)$. Let $a \in \mathbb{R}^m$. Note that $\sum_{s=1}^m a_s(\phi_s \sigma)^\wedge \in \operatorname{Harm}_\mathbb{R}^h(\Omega) \Leftrightarrow \sum_{s=1}^m a_s \phi_s \sigma \in M_\mathbb{R}^h(\partial\Omega) \Leftrightarrow$ (by (1.2.9)) for all $r = 1,\ldots,m$, $0 = \int_{\partial\Omega} \phi_r d \left(\sum_{s=1}^m a_s \phi_s \sigma \right) = \sum_{s=1}^m a_s \int_{\partial\Omega} \phi_r \phi_s d\sigma = a_r \Leftrightarrow a = 0$. $\square$

LEMMA 4.19.

$$(4.20) \qquad \begin{pmatrix} \psi_1 \\ \vdots \\ \psi_m \end{pmatrix} = M_1(\phi^\wedge, \Gamma) \begin{pmatrix} \phi_1 \\ \vdots \\ \phi_m \end{pmatrix}.$$

Furthermore, $\psi_1, \ldots, \psi_m$ are linearly independent and continuous, and satisfy

(4.21) $$\int_{\partial\Omega} \psi_r d\mu = 0 \quad \text{for all } \mu \in M_{\mathbb{R}}^h(\partial\Omega), \ r = 1, \ldots, m.$$

PROOF. Recall $\operatorname{Re} H_\lambda(z) = P_z^\sigma(\lambda) - \sum_{s=1}^m \phi_s(\lambda)(\phi_s\sigma)^\wedge(z)$. Hence for all $\lambda \in \partial\Omega$ and all $r = 1, \ldots, m$,

$$\begin{aligned} 0 &= \int_{\gamma_r} \frac{\partial}{\partial n}\left(P_z^\sigma(\lambda) - \sum_{s=1}^m \phi_s(\lambda)(\phi_s\sigma)^\wedge(z)\right) d\ell(z) \\ &= \int_{\gamma_r} \frac{\partial P_z^\sigma(\lambda)}{\partial n} d\ell(z) - \sum_{s=1}^m \phi_s(\lambda) \int_{\gamma_r} \frac{\partial(\phi_s\sigma)^\wedge}{\partial n} d\ell \\ &= \psi_r(\lambda) - \left(M_1(\phi^\wedge, \Gamma)\begin{pmatrix}\phi_1(\lambda) \\ \vdots \\ \phi_m(\lambda)\end{pmatrix}\right)_r, \end{aligned}$$

implying (4.20). In particular, $\psi_1, \ldots, \psi_m$ are in the linearly span of $\phi_1, \ldots, \phi_m$. Thus the continuity of $\psi_1, \ldots, \psi_m$ follows from the continuity of $\phi_1, \ldots, \phi_m$, and (4.21) follows from (1.2.9). The linearly independence of $\psi_1, \ldots, \psi_m$ follows from (4.20) and the fact that $M_1(\phi^\wedge, \Gamma)$ is nonsingular (Lemma 4.18). $\square$

LEMMA 4.22. Let $u_1, \ldots, u_m \in \operatorname{Harm}_{\mathbb{R}}(\Omega)$. The following are equivalent.

(i) $u_1, \ldots, u_m$ are linearly independent modulo $\operatorname{Harm}_{\mathbb{R}}^h(\Omega)$.
(ii) There exist $\Lambda_1, \ldots, \Lambda_m : \partial\Omega \to \mathbb{R}$ such that for all $\lambda \in \partial\Omega$ (4.12) has a single-valued harmonic conjugate on Ω.

Moreover if (i) and (ii) hold then $\Lambda_1, \ldots, \Lambda_m$ are uniquely determined and are given by (4.14).

PROOF. By the observations preceding the statement of Theorem 4.15, (ii) is equivalent to the following assertion:

(4.23) There exist $\Lambda_1, \ldots, \Lambda_m$ such that (4.13) holds for all $\lambda \in \partial\Omega$.

But by Lemma 4.19, $\psi_1, \ldots, \psi_m$ are linearly independent. Therefore, (4.23) holds if and only if $M_1(u, \Gamma)$ is nonsingular, in which case (4.13) can be solved for $\Lambda_1, \ldots, \Lambda_m$, yielding (4.14). By Lemma 4.17 the assertion that $M_1(u, \Gamma)$ is nonsingular is equivalent to (i). This completes the proof of Lemma 1.6.10. $\square$

PROOF OF THEOREM 4.15. Assume $M_1(u, \Gamma)$ is nonsingular and $\Lambda_1, \ldots, \Lambda_m$ are defined by (4.14). We shall show that (4.16) is the real part of a Schwartz kernel. By Lemma 4.22 we know that for each $\lambda \in \partial\Omega$, (4.16) is the real part of a holomorphic function. Let $k(z, \lambda)$ denote such a holomorphic function, normalized so that

(4.24) $$k(z_0, \lambda) = 0 \quad \text{for all } \lambda \in \partial\Omega$$

(recall that an arbitrary $z_0 \in \Omega$ has been fixed throughout this section). To prove that k is a Schwartz kernel it remains to be shown that for each fixed $z \in \Omega$, $k(z, \lambda)$ is continuous in λ, and that the reproducing property (4.3) holds. We shall make use of the following properties of $\Lambda_1, \ldots, \Lambda_m$, which follow from formula (4.14) and the properties of the ψ_r tabulated in Lemma 4.19:

$$\Lambda_1, \ldots, \Lambda_m \text{ are continuous on } \partial\Omega,$$

and

(4.25) $$\int_{\partial\Omega} \Lambda_s \, d\mu = 0 \quad \text{for all } \mu \in M_{\mathbb{R}}^h(\partial\Omega), \quad s = 1, \ldots, m.$$

To establish the continuity of $k(z, \lambda)$ with respect to λ fix $\lambda_0 \in \partial\Omega$ and suppose the sequence $\{\lambda_n\} \subseteq \partial\Omega$ converges to λ_0. It is well known that $\lambda_n \to \lambda_0$ implies that the sequence of harmonic functions $\{P_z^\sigma(\lambda_n)\}$ converges to $P_z^\sigma(\lambda_0)$ uniformly on compacta. This implies, together with the continuity of $\Lambda_1, \ldots, \Lambda_m$, that

$$\operatorname{Re} k(z, \lambda_n) = P_z^\sigma(\lambda_n) - \sum_{j=1}^{m} \Lambda_j(\lambda_n) u_j(z)$$

$$\to P_z^\sigma(\lambda_0) - \sum_{j=1}^{m} \Lambda_j(\lambda_0) u_j(z)$$

$$= \operatorname{Re} k(z, \lambda_0)$$

uniformly on compacta. Hence, owing to the normalization (4.24), we have

$$k(z, \lambda_n) \to k(z, \lambda_0), \text{ uniformly on compacta.}$$

In particular, $k(z, \lambda_n) \to k(z, \lambda_0)$ for each fixed $z \in \Omega$. This establishes the continuity of $\lambda \mapsto k(z, \lambda)$ at $\lambda = \lambda_0$. Since λ_0 was chosen arbitrarily, $\lambda \mapsto k(z, \lambda)$ must be continuous for all $\lambda \in \partial\Omega$.

We now turn to proving the reproducing property (4.3). First note that for each fixed compact set $K \subseteq \Omega$, $P_z^\sigma(\lambda)$ is bounded on $K \times \partial\Omega$. Thus $\operatorname{Re} k(z, \lambda)$ is bounded on $K \times \partial\Omega$ implying, together with the normalization (4.24), that $k(z, \lambda)$ is bounded on $K \times \partial\Omega$. It follows that for every $\mu \in M_{\mathbb{R}}(\partial\Omega)$ the function

$$g_\mu(z) = \int_{\partial\Omega} k(z, \zeta) \, d\mu(\zeta)$$

is holomorphic in Ω. Let $f \in \operatorname{Hol}_{\text{meas}}(\Omega)$ and let $\mu \in M_{\mathbb{R}}^h(\partial\Omega)$ be the measure induced by $\operatorname{Re} f$. We have

(4.26) $$\operatorname{Re} g_\mu(z) = \int_{\partial\Omega} \operatorname{Re} k(z, \zeta) \, d\mu(\zeta)$$

(4.27) $$= \int_{\partial\Omega} (P_z^\sigma(\zeta) - \sum_{j=1}^{m} \Lambda_j(\zeta) u_j(z)) d\mu(\zeta)$$

(4.28) $$= \int_{\partial\Omega} P_z^\sigma(\zeta) d\mu(\zeta) - \sum_{j=1}^{m} u_j(z) \int_{\partial\Omega} \Lambda_j d\mu$$

(4.29) $$= \mu^\Lambda(z) - 0 \quad \text{(by 4.25)}$$

(4.30) $$= \operatorname{Re} f(z).$$

Therefore, f and g_μ, being holomorphic, can differ by at most an imaginary constant. But (4.24) implies $g_\mu(z_0) = 0$, so we must have

$$f(z) = g_\mu(z) + i \operatorname{Im} f(z_0),$$

proving (4.3).

This completes the proof that if $M_1(u,\Gamma)$ is nonsingular and $\Lambda_1, \ldots, \Lambda_m$ are defined by (4.14) then (4.16) is the real part of a Schwartz kernel.

Conversely assuming that k is a Schwartz kernel on Ω, we shall show that there exists $u = (u_1, \ldots, u_m)$ with each u_r harmonic such that $M_1(u, \Gamma)$ nonsingular and Re k is given by (4.16), where $\Lambda_1, \ldots, \Lambda_m$ satisfy (4.14). By Theorem 4.5, there exists $f_1, \ldots, f_m \in \text{Hol}(\Omega)$ such that

$$k(z, \lambda) = H_\lambda(z) + \sum_{r=1}^m \phi_r(\lambda) f_r(z).$$

Thus

$$\text{Re } k(z, \lambda) = P_z^\sigma(\lambda) - \sum_{r=1}^m \phi_r(\lambda)((\phi_r \sigma)^\wedge(z) - \text{Re } f_r(z)).$$

Let $u_r = (\phi_r \sigma)^\wedge - \text{Re } f_r$ and $\Lambda_r = \phi_r, r = 1, \ldots, m$. That $M_1(u, \Gamma)$ is nonsingular follows from Lemma 4.18 and the fact that $M_1(u, \Gamma) = M_1(\phi^\wedge, \Gamma)$. That $\Lambda_1, \ldots, \Lambda_m$ satisfy (4.14) can be inferred from the uniqueness assertion of Lemma 4.22, or alternatively, verified by direct computation using (4.20). This completes the proof of Theorem 4.15. □

We remark that although the intermediate computations involved in the Coifman–Weiss construction depend on the choice of homology basis Γ, the uniqueness assertion of Lemma 4.22 tells us that the functions $\Lambda_1, \ldots, \Lambda_m$, and hence the resultant Schwartz kernel, are independent of Γ.

An immediate example of the Coifman–Weiss construction results by choosing m harmonic functions $u_1, \ldots, u_m$ which are dual to the homology basis Γ - i.e.,

$$\int_{\gamma_r} \frac{\partial u_s}{\partial n} d\ell = \delta_{rs}, \quad r, s = 1, \ldots, m.$$

In this case $M_1(u, \Gamma)$ is the identity matrix, thus by Theorem 4.15,

$$P_z^\sigma(\lambda) - \sum_{s=1}^m \psi_s(\lambda) u_s(z)$$

is the real part of a Schwartz kernel.

We shall now turn to a discussion of the Khavinson construction, which is better suited than the Coifman–Weiss construction to the special cases in which $u_1, \ldots, u_m$ are presented as boundary data i.e., $u_s = \mu_s^\wedge$, $s = 1, \ldots, m$, where $\mu_1, \ldots, \mu_m \in M_\mathbb{R}(\partial \Omega)$. Writing

$$\mu = (\mu_1, \ldots, \mu_m)$$

and

$$\phi = (\phi_1, \ldots, \phi_m),$$

where $\{\phi_1, \ldots, \phi_m\}$ is our orthonormal basis for $L_\mathbb{R}^2(\sigma) \ominus L_\mathbb{R}^{2,h}(\sigma)$, define the matrix $M_2(\mu, \phi)$ by

$$M_2(\mu, \phi) = \left[\int_{\partial \Omega} \phi_r \, d\mu_s \right]_{r,s=1}^m.$$

As in the case of the Coifman–Weiss construction we attempt to find functions $\Lambda_1, \ldots, \Lambda_m : \partial \Omega \to \mathbb{R}$ such that for each $\lambda \in \partial \Omega$, $z \mapsto P_z^\sigma(\lambda) - \sum_{s=1}^m \Lambda_s(\lambda) \mu_s^\wedge(z)$ has a single-valued harmonic conjugate, or equivalently,

(4.31) $$\delta_\lambda - \sum_{s=1}^{m} \Lambda_s(\lambda)\mu_s \in M_{\mathbb{R}}^h(\partial\Omega) \text{ for all } \lambda \in \partial\Omega.$$

Using (1.2.9) we can attempt to solve for $\Lambda_1, \ldots, \Lambda_m$ explicitly: (4.31) holds if and only if

$$0 = \int_{\partial\Omega} \phi_r \, d(\delta_\lambda - \sum_{s=1}^{m} \Lambda_s(\lambda)\mu_s)$$

$$= \phi_r(\lambda) - \sum_{s=1}^{m} \Lambda_s(\lambda) \int_{\partial\Omega} \phi_r \, d\mu_s \quad \text{for all } \lambda \in \partial\Omega, r = 1, \ldots, m$$

if and only if

$$\begin{pmatrix} \phi_1 \\ \vdots \\ \phi_m \end{pmatrix} = M_2(\mu, \phi) \begin{pmatrix} \Lambda_1 \\ \vdots \\ \Lambda_m \end{pmatrix}.$$

Thus if $M_2(\mu, \phi)$ is nonsingular, (4.31) obtains if and only if

(4.32) $$\begin{pmatrix} \Lambda_1 \\ \vdots \\ \Lambda_m \end{pmatrix} = M_2(\mu, \phi)^{-1} \begin{pmatrix} \phi_1 \\ \vdots \\ \phi_m \end{pmatrix}.$$

THEOREM 4.33. *(The Khavinson construction).* *If $M_2(\mu, \phi)$ is nonsingular and $\Lambda_1, \ldots, \Lambda_m$ are defined by (4.32), then*

(4.34) $$P_z^\sigma(\lambda) - \sum_{s=1}^{m} \Lambda_s(\lambda)\mu_s^\wedge(z)$$

is the real part of a Schwartz kernel k on Ω satisfying

(4.35) $$k(z, \lambda) \in \text{Hol}_{\text{meas}}(\Omega) \text{ for all } \lambda \in \partial\Omega.$$

Conversely, if k is a Schwartz kernel on Ω satisfying (4.35), then there exists an m-tuple of measures $\mu = (\mu_1, \ldots, \mu_m)$ such that $M_2(\mu, \phi)$ is nonsingular and $\text{Re } k(z, \lambda)$ is given by (4.34), where $\Lambda_1, \ldots, \Lambda_m$ satisfy (4.32).

The proof of Theorem 4.33 is analogous to that of Theorem 4.15, and thus will be omitted.

An immediate example of the Khavinson construction results by choosing $\mu_s = \phi_s\sigma$, $s = 1, \ldots, m$. In this case

$$(M_2(\mu, \phi))_{r,s} = \int_{\partial\Omega} \phi_r \, d(\phi_s\sigma) = \int_{\partial\Omega} \phi_r \phi_s d\sigma = \delta_{rs},$$

hence $M_2(\mu, \phi)$ is the identity matrix. Formula (4.32) yields $\Lambda_s = \phi_s$, thus by Theorem 4.33,

$$P_z^\sigma(\lambda) - \sum_{s=1}^{m} \phi_s(\lambda)(\phi_s\sigma)^\wedge(z)$$

is the real part of a Schwartz kernel. Of course, we recognize the above expression as the real part of $H_\lambda(z)$, the Herglotz kernel of the second kind.

The analog of Lemma 4.17 is the statement that the measures $\mu_1, \ldots, \mu_m$ are linearly independent modulo $M_{\mathbb{R}}^h(\partial\Omega)$ if and only if $M_2(\mu, \phi)$ is nonsingular. In specific cases, however, it may be difficult or impossible to compute $M_2(\mu,\phi)$ directly. It is therefore natural to ask whether there is a sufficient criterion for linear independence modulo $M_{\mathbb{R}}^h(\partial\Omega)$ that is more transparent then direct evaluation of $M_2(\mu,\phi)$. The following theorem, a restatement of a result due to D. Khavinson, gives us an elegant sufficient criterion that applies to a large class of sets of positive measures $\{\mu_1,\ldots,\mu_m\} \subseteq M_{\mathbb{R}}(\partial\Omega)$.

THEOREM 4.36. [**Kha84**] *If* $\mu_1,\ldots,\mu_m \in M_{\mathbb{R}}(\partial\Omega)$ *are positive measures satisfying*

(4.37) $$\mu_r(\partial_r) > 0 \quad r = 1,\ldots,m$$

and

(4.38) $$\mu_r(\partial_s) = 0 \quad r = 1,\ldots,m, \ s = 0,\ldots,m, \ r \neq s$$

then $\{\mu_1,\ldots,\mu_m\}$ *are linearly independent modulo* $M_{\mathbb{R}}^h(\partial\Omega)$.

We are now in a position to present and easily justify the two classical examples of Schwartz kernels alluded to earlier in this section. The kernel introduced by Coifman and Weiss in [**CW67**] can be derived by taking

$$\mu = (\chi_{\partial_1}\sigma,\ldots,\chi_{\partial_m}\sigma)$$

where χ_{∂_r} denotes the characteristic function of boundary component ∂_r. By Theorem 4.36, the measures $\chi_{\partial_1}\sigma,\ldots,\chi_{\partial_m}\sigma$ are linearly independent modulo $M_{\mathbb{R}}^h(\partial\Omega)$, thus $M_2(\mu,\phi)$ is nonsingular. Defining $\Lambda_1,\ldots,\Lambda_m$ by (4.32), Theorem 4.33 tells us that

$$P_z^\sigma(\lambda) - \sum_{s=1}^m \Lambda_s(\lambda)(\chi_{\partial_s}\sigma)^\wedge(z)$$

is the real part of a Schwartz kernel on Ω.

Khavinson's kernel, introduced in [**Kha84**], can be arrived at by fixing points $\lambda_r \in \partial_r, r = 1,\ldots,m$, and taking

$$\mu = (\delta_{\lambda_1},\ldots,\delta_{\lambda_m}).$$

Again applying Theorem 4.36, we see that $\delta_{\lambda_1},\ldots,\delta_{\lambda_m}$ are linearly independent modulo $M_{\mathbb{R}}^h(\partial\Omega)$. In this case, $M_2(\mu,\phi)$ can be computed directly:

$$(M_2(\mu,\phi))_{r,s} = \int_{\partial\Omega} \phi_r \, d(\delta_{\lambda_s}) = \phi_r(\lambda_s).$$

Hence by Theorem 4.33,

$$P_z^\sigma(\lambda) - \left\langle \begin{pmatrix} P_z^\sigma(\lambda_1) \\ \vdots \\ P_z^\sigma(\lambda_m) \end{pmatrix}, \left([\phi_r(\lambda_s)]_{r,s=1}^m\right) \begin{pmatrix} \phi_1(\lambda) \\ \vdots \\ \phi_m(\lambda) \end{pmatrix} \right\rangle$$

is the real part of a Schwartz kernel on Ω, where $\langle \cdot, \cdot \rangle$ here denotes the dot product in $\mathbb{R}^m$.

We shall conclude this section by deriving the algebraic relationship between the Coifman–Weiss and Khavinson constructions and by giving the matrix $M_1(\phi^\wedge, \Gamma)$ a Hilbert space interpretation. Let $u_r = \mu_r^\wedge$, $r = 1,\ldots,m$ where $\mu_1,\ldots,\mu_m \in M_{\mathbb{R}}(\partial\Omega)$, and note that since $\mu \mapsto \mu^\wedge$ maps $M_{\mathbb{R}}(\partial\Omega)$ into $\text{Harm}_{\mathbb{R}}^h(\Omega)$ injectively, $u_1,\ldots,u_m$ are linearly independent modulo $\text{Harm}_{\mathbb{R}}^h(\Omega)$ if and only if $\mu_1,\ldots,\mu_m$ are linearly independent modulo $M_{\mathbb{R}}^h(\partial\Omega)$. Thus if $u = (u_1,\ldots,u_m)$ and $\mu =$

4. SCHWARTZ KERNELS ON MULTIPLY CONNECTED DOMAINS

$(\mu_1, \ldots, \mu_m)$, then $M_1(u, \Gamma)$ is nonsingular if and only if $M_2(\mu, \phi)$ is nonsingular. Furthermore, if $M_1(u, \Gamma)$ and $M_2(\mu, \phi)$ are nonsingular then by Theorems 4.15 and 4.33 and the uniqueness assertion of Lemma 4.22, formulas (4.14) and (4.32) must compute the same functions $\Lambda_1, \ldots, \Lambda_m$. A simple and direct algebraic explanation of these connections can be achieved by using the following identity [**For79**]:

$$(4.39) \qquad \int_{\gamma_r} \frac{\partial \mu^\wedge}{\partial n} \, d\ell = \int_{\partial \Omega} \psi_r \, d\mu \quad \text{for all } \mu \in M_\mathbb{R}(\partial \Omega), r = 1, \ldots, m,$$

where $\psi_1, \ldots, \psi_m$ are defined by (1.6.10). This identity follows from the Poisson Formula and Fubini's Theorem. We have

$$M_1(u, \Gamma) = \left[\int_{\gamma_r} \frac{\partial u_s}{\partial n} \, d\ell \right]_{r,s=1}^m$$

$$= \left[\int_{\gamma_r} \frac{\partial \mu_s^\wedge}{\partial n} \, d\ell \right]_{r,s=1}^m$$

$$= \left[\int_{\partial \Omega} \psi_r \, d\mu_s \right]_{r,s=1}^m \quad \text{(by (4.39))}$$

$$(4.40) \qquad = \left[\int_{\partial \Omega} \left(M_1(\phi^\wedge, \Gamma) \begin{pmatrix} \phi_1(\lambda) \\ \vdots \\ \phi_m(\lambda) \end{pmatrix} \right)_r d\mu_s(\lambda) \right]_{r,s=1}^m \quad \text{(by 4.20))}$$

$$= M_1(\phi^\wedge, \Gamma) \left[\int_{\partial \Omega} \phi_t d\mu_s \right]_{t,s=1}^m$$

$$= M_1(\phi^\wedge, \Gamma) M_2(\mu, \phi).$$

Since, by Lemma 4.18, $M_1(\phi^\wedge, \Gamma)$ is nonsingular, we see directly that $M_1(u, \Gamma)$ is nonsingular if and only if $M_2(\mu, \phi)$ is nonsingular. In addition, if $M_1(u, \Gamma)$ and $M_2(\mu, \phi)$ are nonsingular then from (4.40) and (4.20) we obtain

$$M_1(u, \Gamma)^{-1} \begin{pmatrix} \psi_1 \\ \vdots \\ \psi_m \end{pmatrix}$$

$$= (M_1(\phi^\wedge, \Gamma) M_2(\mu, \phi))^{-1} M_1(\phi^\wedge, \Gamma) \begin{pmatrix} \phi_1 \\ \vdots \\ \phi_m \end{pmatrix}$$

$$= M_2(\mu, \phi)^{-1} \begin{pmatrix} \phi_1 \\ \vdots \\ \phi_m \end{pmatrix},$$

thus formulas (4.14) and (4.32) yield identical results.

Note that formula (4.20) is nothing more than the orthonormal expansion of $\psi_1, \ldots, \psi_m$ in terms of $\phi_1, \ldots, \phi_m$ in $L_\mathbb{R}^2(\sigma)$. The conventional way of writing the coefficients in this formula can be recovered with the help of (4.39), which yields

$$(M_1(\phi^\wedge, \Gamma))_{r,s} = \int_{\gamma_r} \frac{\partial(\phi_s \sigma)^\wedge}{\partial n} \, d\ell$$
$$= \int_{\partial \Omega} \psi_r \, d(\phi_s \sigma)$$
$$= \int_{\partial \Omega} \psi_r \phi_s \, d\sigma$$
$$= <\psi_r, \phi_s>_{L^2_{\mathbb{R}}(\sigma)}.$$

APPENDIX A

Convergence Results

In this appendix, we shall establish several regularity and convergence results, which were referred to in the main text. Throughout our theoretical considerations we will make ample use of the following concept.

DEFINITION A.1. *A sequence* $\{a_n\} \subset \mathbb{C}$ *converges to* $a \in \mathbb{C}$ **exponentially fast with factor** $\eta \in (0,1)$ *provided there exists* $c \geq 0$ *such that*

(A.2) $$|a_n - a| \leq c\eta^n \text{ for all } n \geq 1.$$

More simply, we shall also say that $a_n \to a$ exponentially fast if there exists $\eta \in (0,1)$ such that $a_n \to a$ exponentially fast with factor η. Note that if $a_n \to a$ exponentially fast with factor η_1, $b_n \to b$ exponentially fast with factor η_2, and $c_1, c_2 \in \mathbb{C}$, then $c_1 a_n + c_2 b_n \to c_1 a + c_2 b$ exponentially fast with factor $\max\{\eta_1, \eta_2\}$. Also, if $a_n \to 0$ exponentially fast with factor η and p is a polynomial over $\mathbb{C}$ then for all $\epsilon \in (0, 1-\eta)$, $p(n)a_n \to 0$ exponentially fast with factor $\eta + \epsilon$. Finally, if $a_n \to 0$ exponentially fast with factor η then $\sum_{k=n}^{\infty} a_k \xrightarrow{n \to \infty} 0$ exponentially fast with factor η, and $\sqrt{a_n} \to 0$ exponentially fast with factor $\sqrt{\eta}$.

Recall that the measure σ, when restricted to each boundary circle ∂_r, is normalized arc length measure. We shall define an orthogonal basis for $L^2(d\sigma)$ as follows. For $r = 0, \ldots, m$, let e_{-r} denote $\frac{1}{\sqrt{2}}$ times harmonic measure for boundary ∂_r, and for $j \geq 1$ let r, k be such that $f_j = f_{r,k}$, and define $e_j : \partial\Omega \to \mathbb{R}$ by

(A.3) $$e_j(z) = \begin{cases} f_j(z), & \text{if } z \in \partial_r \\ 0, & \text{if } z \in \partial\Omega \setminus \partial_r, \end{cases}$$

Accordingly, if $j = 2l - 1$ then

(A.4) $$e_j(c_r + \rho_r e^{i\theta}) = \cos l\theta,$$

and if $j = 2l$ then

(A.5) $$e_j(c_r + \rho_r e^{i\theta}) = \begin{cases} \sin l\theta, & r = 0 \\ -\sin l\theta, & r \geq 1. \end{cases}$$

Hence, $\{e_j\}_{j=-m}^{\infty}$ is the "standard" orthogonal basis for $L^2(d\sigma)$. (Note that $\|e_j\|^2 = \frac{1}{2}$ for all j, so this basis is not normalized.) This basis was used by Bird and Steele [BS92] to solve Laplace's equation on multiply-connected circle domains. Define

(A.6) $$b_r = \left(\sup_{z \in \partial\Omega \setminus \partial_r} |F_{r,1}(z)|\right)^{\frac{1}{2}} = \begin{cases} \max_{s \geq 1} (|c_s| + \rho_s)^{\frac{1}{2}}, & r = 0 \\ \max_{s \neq r} \left(\frac{\rho_r}{||c_s - c_r| - \rho_s|}\right)^{\frac{1}{2}}, & r = 1, \ldots, m, \end{cases}$$

so that $b_r \in (0,1)$, and for all $l \geq 1$

(A.7) $$\sup_{z \in \partial\Omega \setminus \partial_r} |F_{r,l}(z)| \leq b_r^{2l}$$

and for all $k \geq 1$

(A.8) $$\sup_{z \in \partial\Omega \setminus \partial_r} |f_{r,k}(z)| \leq b_r^{k-1}.$$

Recalling the method of enumeration of $f_1, f_2, \ldots$ given in Section 3.2.1, if $f_j = f_{r,k}$ then
$$N_r\left(\frac{j}{N} - 1\right) \leq k \leq N_r\left(\frac{j}{N} + 1\right).$$

Defining

(A.9) $$b = \max\{b_0^{\frac{N_0}{N}}, \ldots, b_m^{\frac{N_m}{N}}\}$$

and

(A.10) $$x_0 = \max\{b_0^{-N_0-1}, \ldots, b_m^{-N_m-1}\}$$

we see that

(A.11) $$\sup_{z \in \partial\Omega \setminus \partial_r} |f_j(z)| \leq b_r^{k-1}$$

(A.12) $$\leq b_r^{N_r(\frac{j}{N}-1)-1}$$

(A.13) $$\leq x_0 b^j.$$

From (A.3) and (A.13) we obtain

(A.14) $$\sup_{z \in \partial\Omega} |e_j(z) - f_j(z)| \leq x_0 b^j.$$

The next result is an elementary fact from harmonic analysis on the unit circle.

LEMMA A.15. $g \in L^2(\partial\mathbb{D}, \frac{d\theta}{2\pi})$ is real-analytic if an only if

(A.16) $$\int_0^{2\pi} g(e^{i\theta}) \cos n\theta \frac{d\theta}{2\pi} \to 0 \text{ exponentially fast as } n \to \infty$$

and

(A.17) $$\int_0^{2\pi} g(e^{i\theta}) \sin n\theta \frac{d\theta}{2\pi} \to 0 \text{ exponentially fast as } n \to \infty.$$

The above result has the following generalization to $\partial\Omega$.

LEMMA A.18. $g \in L^2_{\mathbb{R}}(d\sigma)$ is real-analytic if and only if
$$\langle g, f_j \rangle \to 0 \text{ exponentially fast as } j \to \infty.$$

PROOF. Let $g \in L^2_{\mathbb{R}}(d\sigma)$. From Lemma A.15, it it clear that g is real-analytic if and only if for all $r = 0, \ldots, m$,

(A.19) $$\int_0^{2\pi} g(c_r + \rho_r e^{i\theta}) \cos(j\theta) \frac{d\theta}{2\pi} \to 0, \text{ exponentially fast as } j \to \infty,$$

and

(A.20) $$\int_0^{2\pi} g(c_r + \rho_r e^{i\theta}) \sin(j\theta) \frac{d\theta}{2\pi} \to 0, \text{ exponentially fast as } j \to \infty,$$

which, by (A.5) is equivalent to the statement that $\langle g, e_j \rangle \to 0$ exponentially fast. But

$$|\langle g, e_j \rangle| \le |\langle g, f_j \rangle| + |\langle g, f_j - e_j \rangle|$$
$$\le |\langle g, f_j \rangle| + \|g\|_2 x_0 b^j \text{ by (A.14)}.$$

Thus, $\langle g, e_j \rangle \to 0$ exponentially fast if and only if $\langle g, f_j \rangle \to 0$ exponentially fast, and the lemma follows. □

PROPOSITION A.21. *If $\varphi \in L^2_{\mathbb{R}}(d\sigma) \ominus L^{2,h}_{\mathbb{R}}(d\sigma)$ then φ is real-analytic.*

PROOF. $\varphi \in L^2_{\mathbb{R}}(d\sigma) \ominus L^{2,h}_{\mathbb{R}}(d\sigma)$ implies that $\langle \varphi, f_j \rangle = 0$ for all j. In particular, $\langle \varphi, f_j \rangle \to 0$ exponentially fast. The result now follows from Lemma A.18 □

The next result establishes the regularity claim made in Section 3.2.2.

PROPOSITION A.22. *If $g : \partial\Omega \to \mathbb{R}$ is real-analytic (resp. C^k, C^∞) then $P_{L^{2,h}(d\sigma)} g$ is real-analytic (resp. C^k, C^∞).*

PROOF. Let $\varphi_1, \ldots, \varphi_m$ be an orthonormal basis for $L^2_{\mathbb{R}}(d\sigma) \ominus L^{2,h}_{\mathbb{R}}(d\sigma)$. By Proposition A.21, $\varphi_1, \ldots, \varphi_m$ are real-analytic. Hence,

$$P_{L^{2,h}_{\mathbb{R}}(d\sigma)} g = g - \sum_{r=1}^m \langle g, \varphi_r \rangle \varphi_r$$

is real-analytic (C^k, C^∞) if g is real-analytic (resp. C^k, C^∞). □

We now turn to the proof of the convergence claims made in sections 3.2.2 - 3.2.5.

LEMMA A.23. *The infinite grammian $G = [\langle f_j, f_i \rangle]_{i,j=0}^\infty$ satisfies:*

(1): $G = \frac{1}{2}I + C$, *where*

(A.24) $$\left(\sum_{k=1}^\infty |C_{jk}|^2 \right)^{\frac{1}{2}} \xrightarrow{j \to \infty} 0 \text{ exponentially fast with factor } b.$$

(2): *There exists $\epsilon > 0$ such that $G \ge \epsilon I$.*

PROOF OF (1). Fix $j \ge 1$ and let r_0 and k_0 be such that

$$f_j = f_{r_0, k_0}.$$

Define for $r, s \in \{0, \ldots, m\}$,

$$S_{r,s}(j) = \left(\sum_{k=1}^\infty |\int_{\partial_s} f_{r,k} f_j d\sigma - \frac{1}{2} \delta_{r_0,s} \delta_{r_0,r} \delta_{k_0,k}|^2 \right)^{\frac{1}{2}}.$$

Let $C = G - \frac{1}{2}I$. We have

$$
\text{(A.25)} \quad \left(\sum_{k=1}^{\infty}|C_{jk}|^2\right)^{\frac{1}{2}} = \left(\sum_{k=1}^{\infty}\left|\left(G - \frac{1}{2}I\right)_{j,k}\right|^2\right)^{\frac{1}{2}}
$$

$$
\text{(A.26)} \quad \leq |\langle f_0, f_j\rangle| + \sum_{s=0}^{m}\sum_{r=0}^{m} S_{r,s}(j).
$$

Now,

$$
\text{(A.27)} \quad |\langle f_0, f_j\rangle| \leq \frac{1}{\sqrt{2(m+1)}}\left(\left|\int_{\partial_{r_0}} f_j d\sigma\right| + \left|\int_{\partial\Omega\setminus\partial_{r_0}} f_j d\sigma\right|\right)
$$

$$
\text{(A.28)} \quad \leq \frac{1}{\sqrt{2(m+1)}}\left(0 + mx_0 b^j\right) \text{ (by (A.13))}
$$

$$
\text{(A.29)} \quad \to 0 \text{ exponentially fast with factor } b.
$$

Hence to prove (A.24), it remains to be shown that for all $r, s \in \{0, \ldots, m\}$,

$$
S_{r,s}(j) \xrightarrow{j \to \infty} 0 \text{ exponentially fast with factor } b.
$$

To do so, we shall consider four cases.

Case 1: $s \neq r$, $s \neq r_0$.
Setting $y_0 = \max_{r \in \{0,\ldots,m\}} \frac{1}{1-b_r}$ we have

$$
S_{r,s}(j) \leq \sum_{k=1}^{\infty} \int_{\partial_s} |f_{r,k} f_j| d\sigma
$$

$$
\leq x_0 b^j \sum_{k=1}^{\infty} b_r^{k-1} \text{ (by (A.8) and (A.13))}
$$

$$
\leq x_0 y_0 b^j.
$$

Case 2: $s = r = r_0$.

$$
S_{r,s}(j) = \left|\int_{\partial_{r_0}} (e_{r_0,k_0})^2 d\sigma - \frac{1}{2}\right|
$$

$$
+ \left(\sum_{\substack{k=1\\k\neq k_0}}^{\infty} \left|\int_{\partial_{r_0}} e_{r_0,k} e_{r_0,k_0} d\sigma\right|^2\right)^{\frac{1}{2}} \text{ (by (A.4) and (A.5))}
$$

$$
= \left|\frac{1}{2} - \frac{1}{2}\right| + 0 = 0.
$$

A. CONVERGENCE RESULTS

Case 3: $s = r \neq r_0$.

$$S_{r,s}(j) = \left(\sum_{k=1}^{\infty} |\int_{\partial_r} e_{r,k} f_j d\sigma|^2\right)^{\frac{1}{2}}$$

$$\leq \frac{1}{\sqrt{2}} \left(\int_{\partial_r} |f_j|^2 d\sigma\right)^{\frac{1}{2}} \text{ by (A.4), (A.5) and Parseval's Theorem.}$$

$$\leq \frac{1}{\sqrt{2}} x_0 b^j \text{ by (A.13).}$$

Case 4: $s = r_0 \neq r$.

Let

$$l_0 = \begin{cases} \frac{k_0+1}{2}, & \text{if } k_0 \text{ is odd} \\ \frac{k_0}{2}, & \text{if } k_0 \text{ is even.} \end{cases}$$

(A.30) $$S_{r,s}(j) = \left(\sum_{l=1}^{\infty} \left(|\int_{\partial_{r_0}} f_{r,2l-1} f_{r_0,k_0} d\sigma|^2 + |\int_{\partial_{r_0}} f_{r,2l} f_{r_0,k_0} d\sigma|^2\right)\right)^{\frac{1}{2}}$$

(A.31) $$\leq \left(\sum_{l=1}^{\infty} |\int_{\partial_{r_0}} F_{r,l} F_{r_0,l_0} d\sigma|^2\right)^{\frac{1}{2}} + \left(\sum_{l=1}^{\infty} |\int_{\partial_{r_0}} F_{r,l} \overline{F_{r_0,l_0}} d\sigma|^2\right)^{\frac{1}{2}}$$

Note that if $z \in \partial_{r_0}$, then

$$\overline{F_{r_0,l_0}(z)} = \frac{1}{F_{r_0,l_0}(z)}$$

For $t = 0, \ldots, m$, let Λ_t denote the parameterized path

(A.32) $$\Lambda_t = c_t + \rho_t e^{i\theta}, \theta \in [0, 2\pi).$$

Recalling that $r \neq r_0$, a change of variables and the residue theorem yield

$$\int_{\partial_{r_0}} F_{r,l} \overline{F_{r_0,l_0}} d\sigma = \int_{\Lambda_{r_0}} \frac{F_{r,l}(z)}{(z - c_{r_0}) F_{r_0,l_0}(z)} \frac{dz}{2\pi i} = 0.$$

Hence,

$$S_{r,s}(j) \leq \left(\sum_{l=1}^{\infty} |\int_{\Lambda_{r_0}} \frac{F_{r,l}(z) F_{r_0,l_0}(z)}{z - c_{r_0}} \frac{dz}{2\pi i}|^2\right)^{\frac{1}{2}}$$

$$= \left(\sum_{l=1}^{\infty} |\int_{\Lambda_r} \frac{F_{r,l}(z) F_{r_0,l_0}(z)}{z - c_{r_0}} \frac{dz}{2\pi i}|^2\right)^{\frac{1}{2}} \text{ by Cauchy's and residue theorems.}$$

$$\leq \left(\sum_{l=-\infty}^{\infty} |\int_0^{2\pi} \frac{F_{r_0,l_0}(c_r + \rho_r e^{i\theta})}{c_r + \rho_r e^{i\theta} - c_{r_0}} \rho_r e^{i\theta} \frac{d\theta}{2\pi}|^2\right)^{\frac{1}{2}}$$

$$= \rho_r \left(\int_0^{2\pi} |\frac{F_{r_0,l_0}(c_r + \rho_r e^{i\theta})}{c_r + \rho_r e^{i\theta} - c_{r_0}}|^2 \frac{d\theta}{2\pi}\right)^{\frac{1}{2}} \text{ by Parseval's Theorem}$$

$$\leq \rho_r x_0 z_0 b^j$$

where

(A.33) $$z_0 = \max\{\frac{1}{b_0^2}, \frac{b_1^2}{\rho_1}, \ldots, \frac{b_m^2}{\rho_m}\}.$$

In summary, we have shown that for all $r, s \in \{0, \ldots, m\}$

$$S_{r,s}(j) \to 0 \text{ exponentially fast with factor } b.$$

This completes the proof of part (1). $\square$

PROOF OF (2). Since G is the weak limit of positive matrices $[\langle f_j, f_i \rangle]_{i,j=0}^N$, $G \geq 0$. To prove part (2), we shall argue by contradiction. Accordingly, assume $0 \in \sigma(G)$, so that $-\frac{1}{2} \in \sigma(C)$. By part (1), C is a compact operator on ℓ^2. By the spectral theorem for self-adjoint compact operators, there exists a nonzero $u \in \ell^2$ such that $Cu = -\frac{1}{2}u$. Hence,

$$Gu = 0.$$

Since the entries of G are real, the entries of u can be taken to be real. Recall the definitions of the functions $\hat{F}_k$ given in (2.5.37), (2.5.38), and (2.5.39). Since for all p, $\max_{z \in \partial \Omega} |\hat{F}_p(z)| \leq 1$, $\sum_{p=0}^\infty u_p \hat{F}_p$ converges in $H^2(\Omega)$, the Hardy space of single-valued analytic functions on Ω with square-integrable boundary values. Let

$$h = \sum_{p=0}^\infty u_p \hat{F}_p,$$

and note that

(A.34) $$\|\mathrm{Re}\, h\|_{L^2(d\sigma)}^2 = \sum_{p,q=0}^\infty u_p u_q \langle f_q, f_p \rangle$$

(A.35) $$= \langle Gu, u \rangle$$

(A.36) $$= 0.$$

Hence, $\mathrm{Re}\, h = 0$ on $\partial \Omega$, and since h is analytic on Ω we have

$$h(z) = ic \text{ for some constant } c \in \mathbb{R}.$$

Define for each $s \in \{0, \ldots, m\}$ and $l \geq 1$ a bounded linear functional $L_{s,l} \in \mathcal{L}(H^2(\Omega))$ by

$$L_{s,l}(g) = \int_{\Lambda_s} \frac{g(z)}{(z - c_s) F_{s,l}(z)} \frac{dz}{2\pi i}.$$

Note that $L_{s,l}(g) = 0$ if g is a constant function. Also, by residue theory, we have for all $r, s \in \{0, \ldots, m\}$ and $l, k \geq 1$,

$$L_{s,l}(\hat{F}_{r,k}) = \begin{cases} 1 & \text{if } k = 2l - 1 \text{ and } r = s, \\ -i, & \text{if } k = 2l \text{ and } r = s, \\ 0, & \text{otherwise}, \end{cases}$$

where $\hat{F}_{r,k}$ is defined in (2.5.40), (2.5.41), and (2.5.42). Recall that for each $j \geq 0$, there is a unique (r, k) such that $\hat{F}_{r,k} = \hat{F}_j$. We shall write $u_j = u_{r,k}$, so that

$$h = \hat{F}_0 + \sum_{r=0}^m \sum_{k=1}^\infty u_{r,k} \hat{F}_{r,k}.$$

We have, for all $s \in \{0, \ldots, m\}$ and $l \geq 1$,

(A.37) $$0 = L_{s,l}(h)$$

(A.38) $$= L_{r,l}(\hat{F}_0) + \sum_{r=0}^{m}\sum_{k=1}^{\infty} u_{r,k} L_{s,l}(\hat{F}_{r,k})$$

(A.39) $$= u_{s,2l-1} - i u_{s,2l},$$

implying, since the $u_{r,k}$ are real, that $u_{s,k} = 0$ for all $s \in \{0, \ldots, m\}$ and $k \geq 1$. But then
$$0 = \operatorname{Re} h = \frac{u_0}{\sqrt{2(m+1)}},$$
implying $u_0 = 0$ also. Therefore, $u_p = 0$ for all $p \geq 0$. This contradicts $u \neq 0$, and hence we conclude that $0 \notin \sigma(G)$, and the proof of part(2) is complete. $\square$

The following definition formalizes the concept of exponential convergence in C^∞ introduced in Section 3.2.2.

DEFINITION A.40. $\{g_n\} \subseteq C^\infty$ is **exponentially Cauchy** in C^∞ if there exists $\eta \in (0, 1)$ such that for all $k \geq 0$

(A.41) $$\max_{\lambda \in \partial \Omega} |g_n^{(k)}(\lambda) - g_{n-1}^{(k)}(\lambda)| \xrightarrow{n \to \infty} 0 \text{ exponentially fast with factor } \eta.$$

It is straightforward to show that if $\{g_n\}$ is exponentially Cauchy in C^∞ then there exists $g \in C^\infty$ such that $g_n \to g$ exponentially fast in C^∞; moreover, if $\{g_n\}$ is exponentially Cauchy and there exists $g \in L^2(d\sigma)$ such that $g_n \to g$ in $L^2(d\sigma)$ then $g \in C^\infty$ and $g_n \to g$ exponentially fast in C^∞. Therefore, the proving that $P_{\mathcal{M}_N} g \to P_{L^{2,h}(d\sigma)} g$ exponentially fast in C^∞ reduces to showing that $\{P_{\mathcal{M}_N} g\}_{N=1}^{\infty}$ is exponentially Cauchy in C^∞. The next result furnishes a key estimate that we shall employ to prove that various sequences are exponentially Cauchy in C^∞. For the purposes of facilitating the statement of this result, we define real constants d_1, d_2, and d_3 as follows. Recall, (A.9) and (A.10), and set
$$d_1 = \frac{x_0}{\sqrt{1-b^2}} + \frac{1}{\sqrt{2}}.$$
By Lemma A.23(1), we know that there exists a constant d_2 satisfying
$$\left(\sum_{j=1}^{\infty}(G_{N+1,j} - \frac{1}{2}\delta_{N+1,j})^2\right)^{\frac{1}{2}} \leq d_2 b^N \text{ for all } N \geq 0,$$
and by Lemma A.23(2), the constant d_3 defined by
$$d_3 = \max\{\|G^{-1}\|, 2\}$$
is finite.

LEMMA A.42. For every $k \geq 0$ there exists a polynomial p_k of degree k or less with nonnegative coefficients such that for all $g \in L^2(d\sigma)$
$$\sup_{\lambda \in \partial\Omega} |(P_{\mathcal{M}_N}g)^{(k)}(\lambda) - (P_{\mathcal{M}_{N-1}}g)^{(k)}(\lambda)|$$
$$\leq \sqrt{2} d_1 d_2 d_3^2 \sqrt{N+1} b^N p_k(N) \|g\| + 2|\langle g, f_N \rangle| p_k(N).$$

PROOF. Fix $g \in L^2(d\sigma)$. By (2.5.35)

$$\left(P_{\mathcal{M}_N}g\right)^{(k)}(\lambda) - \left(P_{\mathcal{M}_{N-1}}g\right)^{(k)}(\lambda) = \langle g, \mathbf{f_N}\rangle \left(G_N^{-1} - \begin{pmatrix} G_{N-1}^{-1} & 0 \\ 0 & 0 \end{pmatrix}\right) \begin{pmatrix} f_0^{(k)}(\lambda) \\ \cdot \\ \cdot \\ \cdot \\ f_N^{(k)}(\lambda) \end{pmatrix}.$$

We shall establish a number of inequalities that will be used to bound the above expression. Recalling (A.14) and the fact that $\{\sqrt{2}e_j\}_{j=-m}^{\infty}$ is an orthonormal basis for $L^2(d\sigma)$ we have

(A.43) $\qquad \|\langle g, \mathbf{f_N}\rangle\|_2 = \|(\langle g, f_0\rangle, \langle g, f_1\rangle, \ldots, \langle g, f_N\rangle)\|_2$

(A.44) $\qquad \leq \left(\sum_{j=0}^{N} |\langle g, f_j - e_j\rangle|^2\right)^{\frac{1}{2}} + \left(\sum_{j=0}^{N} |\langle g, e_j\rangle|^2\right)^{\frac{1}{2}}$

(A.45) $\qquad \leq \|g\| \left(\sum_{j=0}^{\infty} \|f_j - e_j\|^2\right)^{\frac{1}{2}} + \frac{1}{\sqrt{2}} \left(\sum_{j=-m}^{\infty} |\langle g, \sqrt{2}e_j\rangle|^2\right)^{\frac{1}{2}}$

(A.46) $\qquad \leq \frac{x_0}{\sqrt{1-b^2}} \|g\| + \frac{1}{\sqrt{2}} \|g\|$

(A.47) $\qquad \leq d_1 \|g\|.$

From the definitions of d_2 and d_3 we have

(A.48) $\quad \left\| G_N^{-1} - \begin{pmatrix} G_{N-1}^{-1} & 0 \\ 0 & 2 \end{pmatrix} \right\| = \left\| G_N^{-1} \left(\begin{pmatrix} G_{N-1} & 0 \\ 0 & \frac{1}{2} \end{pmatrix} - G_N\right) \begin{pmatrix} G_{N-1}^{-1} & 0 \\ 0 & 2 \end{pmatrix} \right\|$

(A.49) $\qquad \leq d_3^2 \left(2\sum_{j=1}^{N} (G_{N+1,j})^2 + (G_{N+1,N+1} - \frac{1}{2})^2\right)^{\frac{1}{2}}$

(A.50) $\qquad \leq \sqrt{2}d_3^2 \left(\sum_{j=1}^{\infty} (G_{N+1,j} - \frac{1}{2}\delta_{N+1,j})^2\right)^{\frac{1}{2}}$

(A.51) $\qquad \leq \sqrt{2}d_2 d_3^2 b^N.$

Next, note that if $\lambda \in \partial_s$ then $\lambda = c_s + \rho_s e^{i\theta}$ for some $\theta \in [0, 2\pi)$, and

(A.52) $\qquad |f_{r,2l-1}^{(k)}(\lambda)| = |\text{Re}\frac{d^k}{d\theta^k} F_{r,l}(c_s + \rho_s e^{i\theta})|$

(A.53) $\qquad \leq |\frac{d^k}{d\theta^k} F_{r,l}(c_s + \rho_s e^{i\theta})|.$

Similarly,

$$|f_{r,2l}^{(k)}(\lambda)| \leq |\frac{d^k}{d\theta^k} F_{r,l}(c_s + \rho_s e^{i\theta})|.$$

Now, by direct computation, if $r = 0$ then

$$\frac{d}{d\theta} F_{r,l}(c_s + \rho_s e^{i\theta}) = il\left[F_{0,l}(c_s + \rho_s e^{i\theta}) - c_s F_{0,l-1}(c_s + \rho_s e^{i\theta})\right],$$

and if $r \geq 1$,
$$\frac{d}{d\theta}F_{r,l}(c_s + \rho_s e^{i\theta}) = -il\left[F_{r,l}(c_s + \rho_s e^{i\theta}) + \frac{c_r - c_s}{\rho_r}F_{r,l+1}(c_s + \rho_s e^{i\theta})\right].$$

By applying induction to the above recursive formulae, it is straightforward to show that there exists a polynomial p_k with non-negative coefficients and degree$(p_k) \leq k$ such that

(A.54)
$$\max_{\substack{\lambda \in \partial\Omega \\ 0 \leq l \leq k}} |f_j^{(l)}(\lambda)| \leq p_k(j).$$

Thus,
$$\sup_{\lambda \in \partial\Omega} \left\| \begin{pmatrix} f_0^{(k)}(\lambda) \\ \vdots \\ f_N^{(k)}(\lambda) \end{pmatrix} \right\|_2 \leq \sqrt{N+1}\, p_k(N).$$

Combining these estimates, we have
$$\sup_{\lambda \in \partial\Omega} |(P_{\mathcal{M}_N}g)^{(k)}(\lambda) - (P_{\mathcal{M}_{N-1}}g)^{(k)}(\lambda)|$$

$$\leq \max_{\lambda \in \partial\Omega} |\langle g, \mathbf{f_N}\rangle| \left(G_N^{-1} - \begin{pmatrix} G_{N-1}^{-1} & 0 \\ 0 & 0 \end{pmatrix} \right) \begin{pmatrix} f_0^{(k)}(\lambda) \\ \vdots \\ f_N^{(k)}(\lambda) \end{pmatrix} |$$

$$\leq \max_{\lambda \in \partial\Omega} |\langle g, \mathbf{f_N}\rangle| \left(G_N^{-1} - \begin{pmatrix} G_{N-1}^{-1} & 0 \\ 0 & 2 \end{pmatrix} \right) \begin{pmatrix} f_0^{(k)}(\lambda) \\ \vdots \\ f_N^{(k)}(\lambda) \end{pmatrix} |$$

$$+ \max_{\lambda \in \partial\Omega} |\langle g, \mathbf{f_N}\rangle| \left(\begin{pmatrix} G_{N-1}^{-1} & 0 \\ 0 & 2 \end{pmatrix} - \begin{pmatrix} G_{N-1}^{-1} & 0 \\ 0 & 0 \end{pmatrix} \right) \begin{pmatrix} f_0^{(k)}(\lambda) \\ \vdots \\ f_N^{(k)}(\lambda) \end{pmatrix} |$$

$$\leq d_1 \|g\| \sqrt{2 d_2 d_3} b^N \sqrt{N+1}\, p_k(N) + 2|\langle g, f_N\rangle| p_k(N).$$

This concludes the proof of Lemma A.42. $\square$

PROPOSITION A.55. *If $g : \partial\Omega \to \mathbb{R}$ is real-analytic, then $P_{\mathcal{M}_n}g \to P_{L^{2,h}(d\sigma)}g$ exponentially fast in C^∞.*

PROOF. Fix a real-analytic $g : \partial\Omega \to \mathbb{R}$. By Lemma A.18, there exists $\eta_0 \in (0,1)$ such that $|\langle g, f_N\rangle| \to 0$ exponentially fast with factor η_0. Let
$$\epsilon = \frac{1}{2}(1 - \max(\eta_0, b))$$
and note that by Lemma A.42 and the remarks following Definition A.1
$$\max_{\lambda \in \partial\Omega} |(P_{\mathcal{M}_N}g)^{(k)}(\lambda) - (P_{\mathcal{M}_{N-1}}g)^{(k)}(\lambda)| \to 0 \text{ exponentially fast with factor } 1 - \epsilon.$$

Thus, $\{P_{\mathcal{M}_N}g\}$ is exponentially Cauchy in C^∞. □

We shall now define a particular linear and continuous harmonic conjugation operator $T: L^{2,h}(d\sigma) \to L^{2,h}(d\sigma)$ as follows. For $a \in \mathbb{R}^{2n+1}$ and $\mathbf{f_{2n}} = (f_0, \ldots, f_{2n})$ let $\langle a, \mathbf{f_{2n}} \rangle = \sum_{j=0}^{2n} a_j f_j$. Define

$$S = \bigcup_{n=1}^{\infty} \mathcal{M}_n = \{\langle \mathbf{a}, \mathbf{f_{2n}} \rangle : \mathbf{a} \in \mathbb{R}^{2n+1}, n \geq 1\}.$$

Note that S is dense in $L^{2,h}$. For each $n \geq 1$ define the matrix $U_n \in \mathcal{L}(\mathbb{R}^{2n+1})$ by

$$U_n = (0) \bigoplus_{j=1}^{n} \begin{pmatrix} 0 & -1 \\ 1 & 0 \end{pmatrix}.$$

Since for all $k \geq 1$, f_{2k} is a harmonic conjugate of f_{2k-1} and $-f_{2k-1}$ is a harmonic conjugate of f_{2k}, for all $a \in \mathbb{R}^{2n+1}$ and all $n \geq 1$, $\langle U_n \mathbf{a}, \mathbf{f_{2n}} \rangle$ is a harmonic conjugate of $\langle a, f_{2n} \rangle$.

DEFINITION A.56. *Let* $\mathbf{f_{2n}} = (f_0, f_1, \ldots, f_{2n})$. *For* $\mathbf{a} = (a_0, a_1, \ldots, a_{2n}) \in \mathbb{R}^{2m+1}$, *define a harmonic conjugation operator* $T: S \to S$ *by*

$$T(\langle \mathbf{a}, \mathbf{f_{2n}} \rangle) = \langle U_n \mathbf{a}, \mathbf{f_{2n}} \rangle.$$

Recalling the definition of $\widehat{F}_j$ in (2.5.37) - (2.5.39), note that for all $j \geq 0$

$$f_j + iT(f_j) = \hat{F}_j.$$

Furthermore, T is well-defined on all of S since the functions f_k are linearly independent. Also, T is evidently linear. To establish that T is bounded one must take into account that the functions f_j are not normalized in the conventional sense - that is, it is not the case that $f_j(z_0) = 0$ for some fixed $z_0 \in \Omega$. With this conventional normalization for the harmonic conjugate, (i.e. $f(z_0) = 0$) it is well known that harmonic conjugation is a contraction of $L^2_{\mathbb{R}}(\omega_{z_0})$, where ω_{z_0} is harmonic measure on $\partial\Omega$ for the point z_0. Since ω_{z_0} is mutually absolutely continuous with respect to normalized arc length measure, σ, this fact together with the inequality $|f(z_0)| \leq \|P^\sigma_{z_0}\|_{L^2(\sigma)} \|f\|_{L^2(\sigma)}$ can be used to prove that T is bounded. Alternatively, we can establish the boundedness of T by direct computation:

$$\text{(A.57)} \qquad \frac{\|T\langle \mathbf{a}, \mathbf{f_{2n}} \rangle\|^2}{\|\langle \mathbf{a}, \mathbf{f_{2n}} \rangle\|^2} = \frac{\sum_{i,j=0}^{2n}(U_n\mathbf{a})_i(U_n\mathbf{a})_j \langle f_j, f_i \rangle}{\sum_{i,j=0}^{2n} a_i a_j \langle f_j, f_i \rangle}$$

$$\text{(A.58)} \qquad = \frac{\langle G_{2n} U_n \mathbf{a}, U_n \mathbf{a} \rangle}{\langle G_{2n} \mathbf{a}, \mathbf{a} \rangle}$$

$$\text{(A.59)} \qquad = \frac{\langle U_n^* G_{2n} U_n \mathbf{a}, \mathbf{a} \rangle}{\langle G_{2n} \mathbf{a}, \mathbf{a} \rangle}.$$

Since $\|U_n^*\| = \|U_n\| = 1$, we have

$$\text{(A.60)} \qquad \langle U_n^* G_{2n} U_n \mathbf{a}, \mathbf{a} \rangle \leq \|U_n^* G_{2n} U_n\| \|\mathbf{a}\|^2$$

$$\text{(A.61)} \qquad \leq \|G_{2n}\| \|\mathbf{a}\|^2$$

$$\text{(A.62)} \qquad \leq \|G\| \|\mathbf{a}\|^2.$$

Recalling Lemma A.23(2), there exists $\epsilon > 0$ such that $\langle G_{2n}\mathbf{a}, \mathbf{a}\rangle \geq \epsilon \|\mathbf{a}\|^2$, and we conclude that
$$\|T\|^2 \leq \frac{1}{\epsilon}\|G\|.$$
It follows that T extends to a bounded linear operator on $L^{2,h}(d\sigma)$. Since T is a harmonic conjugation operator on the dense set $S \subseteq L^{2,h}(d\sigma)$, T is a harmonic conjugation operator on $L^{2,h}(d\sigma)$.

LEMMA A.63. *For every $k \geq 0$, there exists a polynomial p_k of degree k or less with nonnegative coefficients such that for all $g \in L^{2,h}(d\sigma)$,*
$$\sup_{\lambda \in \partial\Omega} |(TP_{\mathcal{M}_N}g)^{(k)}(\lambda) - (TP_{\mathcal{M}_{N-1}}g)^{(k)}(\lambda)|$$
$$\leq \sqrt{2}d_1 d_2 d_3^2 \sqrt{N+1} b^N p_k(N+1)\|g\| + 2\langle g, f_N\rangle p_k(N+1).$$

PROOF. Note that
$$(TP_{\mathcal{M}_n}g)^{(k)} = \langle g, \mathbf{f_n}\rangle G_n^{-1} \begin{pmatrix} (Tf_0)^{(k)} \\ \cdot \\ \cdot \\ \cdot \\ (Tf_n)^{(k)} \end{pmatrix}.$$

Since for $k \geq 1$
(A.64) $$T(f_{2k-1}) = f_{2k}$$
and
(A.65) $$T(f_{2k}) = -f_{2k-1}$$
or equivalently, for all $r = 0, \ldots, m$, $l \geq 1$
(A.66) $$T(f_{r,2l-1}) = f_{r,2l}$$
and
(A.67) $$T(f_{r,2l}) = -f_{r,2l-1},$$
it follows from (A.54) that
(A.68) $$\max_{\substack{\lambda \in \partial\Omega \\ 0 \leq l \leq k}} |(Tf_j)^{(l)}(\lambda)| \leq p_k(j+1).$$

The proof of the lemma proceeds exactly as the proof of Lemma A.42 with
$$\left((Tf_0)^{(k)}(\lambda) \ldots (Tf_n)^{(k)}(\lambda)\right)^t$$
replacing $\left(f_0^{(k)}(\lambda) \ldots f_n^{(k)}(\lambda)\right)^t$ and with (A.68) replacing (A.54). □

We shall now establish that the sequences $\{P_{\mathcal{M}_N}Q_\lambda\}$ and $\{TP_{\mathcal{M}_N}Q_\lambda\}$ converge not only exponentially fast in C^∞, but also uniformly in λ. This result has significance with regards to the convergence of $H_\lambda^{\mathbf{N}}$, as discussed in Section 3.2.4. The proof of uniformity hinges on the following estimate.

LEMMA A.69. *If b is defined as in (A.9), then there exists $a > 0$ such that*
(A.70) $$|\langle Q_\lambda^r, f_j\rangle| \leq ab^j \text{ for all } r \in \{0, \ldots, m\}, \lambda \in \partial_r, j \geq 0.$$

PROOF. Fix r and $\lambda \in \partial_r$. Note that for all $z \in \partial\Omega \setminus \partial_r$

(A.71) $$|Q^r_\lambda(z)| \leq |\frac{\lambda + z - 2c_r}{z - \lambda}|$$

(A.72) $$\leq a_0,$$

where

(A.73) $$a_0 = \frac{2}{\min_{r,s \in \{0,\ldots,m\}, r \neq s} \operatorname{dist}(\partial_r, \partial_s)}$$

(A.74) $$= \frac{2}{\min\{\min_{r,s \geq 1, r \neq s}(|c_s - c_r| - \rho_s - \rho_r), \min_{r \geq 1}(1 - |c_r| - \rho_r)\}}.$$

Since $Q^r_\lambda(z) \equiv 0$ for $z \in \partial_r$, it is clear from (A.13) and (A.72) that if $f_j = f_{s,k}$, where $s = r$, then $|\langle Q^r_\lambda, f_j \rangle| \leq ab^j$ holds with $a = a_0 x_0 m$. On the other hand, if $f_j = f_{s,k}$ and $s \neq r$, then

(A.75) $$|\langle Q^r_\lambda, f_j \rangle| \leq |\int_{\partial_r} Q^r_\lambda f_j d\sigma| + |\int_{\partial_s} Q^r_\lambda f_j d\sigma| + \sum_{\substack{t=0 \\ t \neq r, t \neq s}}^{m} |\int_{\partial_t} Q^r_\lambda f_j d\sigma|$$

(A.76) $$= 0 + |\int_{\partial_s} Q^r_\lambda f_j d\sigma| + (m-1)a_0 x_0 b^j.$$

Hence, the proof will be complete once we establish that there exists a constant a_1 such that for all $s \in \{0, \ldots m\}$, all $\lambda \in \partial_r$, and all $f_j = f_{s,k}$ such that $s \neq r$,

(A.77) $$|\int_{\partial_s} Q^r_\lambda f_j d\sigma| \leq a_1 b^j.$$

To establish (A.77), fix $r \in \{0, \ldots, m\}$ and $\lambda \in \partial_r$ and let $f_j = f_{s,k}$ where $s \neq r$. Let

$$l = \begin{cases} \frac{k}{2} & \text{if } k \text{ is even,} \\ \frac{k+1}{2} & \text{if } k \text{ is odd.} \end{cases}$$

Without loss of generality, we can assume $k \geq 4$, so that $l \geq 2$. By the definition of $F_{s,l}$, if $z \in \partial_s$, then

$$\overline{F_{s,l}(z)} = \frac{1}{F_{s,l}(z)}.$$

Recalling (A.32), we have

(A.78) $$|\int_{\partial_s} Q^r_\lambda f_j d\sigma| \leq \frac{1}{2}\Big(|\int_{\Lambda_s} \frac{\lambda + z - 2c_r}{z - \lambda} \frac{F_{s,l}(z)}{z - c_s} \frac{dz}{2\pi i}|$$

(A.79) $$+ |\int_{\Lambda_s} \frac{\lambda + z - 2c_r}{z - \lambda} \frac{1}{F_{s,l}(z)(z - c_s)} \frac{dz}{2\pi i}|\Big)$$

(A.80) $$= |\frac{\lambda - c_r}{\lambda - c_s} F_{s,l}(\lambda)| + 0 \text{ by residue theory}$$

(A.81) $$\leq \rho_r x_0 z_0 b^j,$$

where x_0, z_0, and b are defined by (A.10), (A.33), and (A.9), respectively. Hence, since $\rho_r \leq 1$, we have established (A.77) with $a_1 = x_0 z_0$. This completes the proof of the lemma. □

PROPOSITION A.82. *Let Q^r_λ be defined as in (3.2.21), and let $b \in \mathbb{R}$ be defined as in (A.9).*

(1) For every $k \geq 0$ and every $\epsilon \in (0, 1-b)$, there exists a constant $d_k^1 \geq 0$ such that for all $r \in \{0, \ldots, m\}$ and all $\lambda \in \partial_r$

(A.83) $$\sup_{z \in \partial\Omega} |(P_{\mathcal{M}_N} Q_\lambda^r)^{(k)}(z) - (P_{L^{2,h}(d\sigma)} Q_\lambda^r)^{(k)}(z)| \leq d_k^1 (b+\epsilon)^N.$$

(2) For every $k \geq 0$ and every $\epsilon \in (0, 1-b)$, there exists $d_k^2 \geq 0$ such that for all $r \in \{0, \ldots, m\}$ and all $\lambda \in \partial_r$

(A.84) $$\sup_{z \in \partial\Omega} |(TP_{\mathcal{M}_N} Q_\lambda^r)^{(k)}(z) - (TP_{L^{2,h}(d\sigma)} Q_\lambda^r)^{(k)}(z)| \leq d_k^2 (b+\epsilon)^N.$$

PROOF. By (A.72), $\|Q_\lambda^r\|_2 \leq a_0$ for all r and λ. Thus, by Lemmas A.42 and A.69, for every $k \geq 0$ and every $\epsilon \in (0, 1-b)$, there exists $a_k \geq 0$ such that for all $r \in \{0, \ldots, m\}$ and all $\lambda \in \partial_r$

$$\sup_{z \in \partial\Omega} |(P_{\mathcal{M}_N} Q_\lambda^r)^{(k)}(z) - (P_{\mathcal{M}_{N-1}} Q_\lambda^r)^{(k)}(z)| \leq a_k (b+\epsilon)^N.$$

Also, for all $g \in L^2(\sigma)$,

$$P_{L^{2,h}(d\sigma)} g - P_{\mathcal{M}_N} g = \sum_{j=N+1}^{\infty} (P_{\mathcal{M}_j} g - P_{\mathcal{M}_{j-1}} g),$$

and by Proposition A.55 if g is real-analytic then the series converges in C^∞. Consequently,

$$\sup_{z \in \partial\Omega} |(P_{L^{2,h}(d\sigma)} Q_\lambda^r)^{(k)}(z) - (P_{\mathcal{M}_n} Q_\lambda^r)^{(k)}(z)|$$
$$\leq \sum_{j=N+1}^{\infty} \sup_{z \in \partial\Omega} |(P_{\mathcal{M}_j} Q_\lambda^r)^{(k)}(z) - (P_{\mathcal{M}_{j-1}} Q_\lambda^r)^{(k)}(z)|$$
$$\leq \sum_{j=N+1}^{\infty} a_k (b+\epsilon)^j$$
$$= \frac{a_k (b+\epsilon)^{N+1}}{1-b-\epsilon}.$$

Setting $d_k^1 = \frac{a_k(b+\epsilon)}{1-b-\epsilon}$, this completes the proof of part(1).

To prove part (2), first note that by Lemma A.69 and formulae (A.64)-(A.65),

$$|\langle Q_\lambda^r, Tf_j \rangle| \leq ab^{j-1} \text{ for all } r \in \{0, \ldots, m\}, \lambda \in \partial_r, j \geq 0.$$

Using Lemma A.63, the proof of part (2) is now identical to the proof of part (1). □

The remainder of this section is devoted to characterizing the error function $P_{\mathcal{M}_N} g - P_{L^{2,h}(d\sigma)} g$ when g in $L^2(d\sigma)$. We begin with a definition.

Let $E_N = [e_{-m}, e_{-m+1}, \ldots, e_0, \ldots, e_N]$.

DEFINITION A.85. $\{g_n\} \subset L^2(d\sigma)$ is **oscillatory** if there exists a sequence $\{a_n\} \subseteq \mathbb{R}$ such that $a_n \to 0$ exponentially fast and

$$\|P_{E_N} g_N\| \leq |a_N| \|g_N\| \text{ for all } N.$$

Note that if $\{g_n\}$ is oscillatory, then for large N, either $g_N = 0$ or
$$g_N = \sum_{k=N+1}^{\infty} c_n e_n + \epsilon_N,$$
where $\|\epsilon_N\| \ll \|g_N\|$, i.e., g is constituted primarily of "high frequency" terms. Below we shall show that if $g \in L^{2,h}(d\sigma)$, then $\{P_{L^{2,h}(d\sigma)} - P_{\mathcal{M}_N} g\}$ is oscillatory. First, we establish the following lemma.

LEMMA A.86.
$$\|P_{L^{2,h}(d\sigma)\ominus\mathcal{M}_N} P_{E_N}\| \to 0 \text{ exponentially fast.}$$

PROOF. We claim, and shall prove below, that there exists $a_0 > 0$ such that for all $N \geq 1$ and all $j \leq N$,

(A.87) $$\|P_{L^{2,h}(d\sigma)\ominus\mathcal{M}_N} e_j\| \leq a_0 \sqrt{N} b^N,$$

where b is given by (A.9). Assuming (A.87) for the moment, and recalling that $\{\sqrt{2} e_j\}$ is an orthonormal basis for $L^2(d\sigma)$, we have

(A.88) $$\|P_{L^{2,h}(d\sigma)\ominus\mathcal{M}_N} P_{E_N}\| = \sup_{\|x\|=1} \|(P_{L^{2,h}(d\sigma)\ominus\mathcal{M}_N} P_{E_N})x\|$$

(A.89) $$= \sup_{\|x\|=1} \|P_{L^{2,h}(d\sigma)\ominus\mathcal{M}_N}(2 \sum_{j=-m}^{N} \langle x, e_j \rangle e_j)\|$$

(A.90) $$\leq \sqrt{2} \sum_{j=-m}^{N} \|P_{L^{2,h}(d\sigma)\ominus\mathcal{M}_N} e_j\|$$

(A.91) $$\leq \sqrt{2} a_0 \sqrt{N}(N+m+1) b^N \text{ by (A.87)},$$

which implies the lemma. We shall now establish (A.87). Note that if $x \in L^2(d\sigma)$, then
$$P_{L^{2,h}(d\sigma)\ominus\mathcal{M}_N} x = \sum_{l=0}^{\infty} \left(P_{\mathcal{M}_{N+l+1}} x - P_{\mathcal{M}_{N+l}} x \right).$$

Thus, it suffices to show that there exists $a_1 > 0$ such that for all $l \geq 0$ and all $j \leq N$

(A.92) $$\|P_{\mathcal{M}_{N+l+1}} e_j - P_{\mathcal{M}_{N+l}} e_j\| \leq a_1 b^{N+l+1} \sqrt{N+l+2}.$$

To prove (A.92), first note that by (A.13) and by the definition of e_j and f_j

(A.93) $$|\langle e_j, f_t \rangle| \leq x_0 b^t \text{ if } t \neq j.$$

Invoking Lemma A.42 with $k = 0$ (so that $p_0(N) \equiv p_0 \geq 0$), we have, for all $j \leq N$
$$\sup_{\lambda \in \partial \Omega} |P_{\mathcal{M}_{N+l+1}} e_j(\lambda) - P_{\mathcal{M}_{N+l}} e_j(\lambda)|$$
$$\leq \sqrt{2} d_1 d_2 d_3^2 \sqrt{N+l+2} b^{N+l+1} p_0 \|e_j\| + 2|\langle e_j, f_{N+l+1}\rangle| p_0.$$

Together with (A.93), the above inequality implies (A.92) upon setting
$$a_1 = (d_1 d_2 d_3^2 + 2x_0) p_0.$$
□

PROPOSITION A.94. If $g \in L^2(d\sigma)$ then $\{P_{L^{2,h}(d\sigma)} g - P_{\mathcal{M}_N} g\}$ is oscillatory.

PROOF. Fix $g \in L^2_{\mathbb{R}}(d\sigma)$. Let $g_N = P_{L^{2,h}(d\sigma)}g - P_{\mathcal{M}_N}g = P_{L^{2,h}(d\sigma)\ominus\mathcal{M}_N}g$. Because orthogonal projections are self-adjoint idempotents, we have

(A.95) $$\|P_{E_N}g_N\|^2 = \langle P_{E_N}g_N, P_{E_N}g_N\rangle$$
(A.96) $$= \langle P_{E_N}g_N, g_N\rangle$$
(A.97) $$= \langle P_{L^{2,h}(d\sigma)\ominus\mathcal{M}_N}P_{E_N}g_N, g_N\rangle$$
(A.98) $$\leq \|P_{L^{2,h}(d\sigma)\ominus\mathcal{M}_N}P_{E_N}\|\|g_N\|^2.$$

The proof is now completed by invoking Lemma A.86. □

APPENDIX B

Example Inner Product Computation

Suppose we wish to compute $\langle f_j, f_i \rangle$ where $f_j = \text{Re} F, f_i = \text{Re} H$ for $F, H \in Rat(\Omega^-)$. According to the definition of the inner product on $L^2(d\sigma)$ we have

(B.1) $$\langle f_j, f_i \rangle = \frac{1}{4} \sum_{r=0}^{m} \int_{\partial_r} \left(F(\lambda) + \overline{F(\lambda)} \right) \left(H(\lambda) + \overline{H(\lambda)} \right) d\sigma_r(\lambda).$$

We expand the integrand into a sum of four terms. Here we demonstrate the computation of one of these terms: $\int_{\partial_r} \overline{F(\lambda)} H(\lambda) d\sigma_r(\lambda)$; the other terms are computed analogously.

Since $F \in Rat(\Omega^-)$, $\overline{F(\lambda)}$ is a rational function of $\overline{\lambda}$. Let us denote this function by $\tilde{F}(\overline{\lambda})$. If $\lambda \in \partial_r$, then $\lambda = \rho_r e^{i\theta} + c_r$ and $d\sigma_r(\lambda) = \frac{d\theta}{2\pi}$. If we set $z = e^{i\theta}$, then $\overline{\lambda} = \frac{\rho_r}{z} + \overline{c_r}$, and $dz = iz d\theta$. Thus

(B.2) $$\int_{\partial_r} \overline{F} H d\sigma_r(\lambda) = \int_{\partial \mathbb{D}} \tilde{F}(\frac{\rho_r}{z} + \overline{c_r}) H(\rho_r z + c_r) \frac{dz}{2\pi i z}.$$

Let $L(z) = \frac{1}{z} \tilde{F}(\frac{\rho_r}{z} + \overline{c_r}) H(\rho_r z + c_r)$, and let $\mathcal{P}$ be the set of poles of $L(z)$. Residue theory allows us to compute the contour integral in equation (B.2) via

(B.3) $$\int_{\partial_r} \overline{F} H d\sigma_r(\lambda) = \sum_{z_0 \in (\mathcal{P} \cap \mathbb{D})} Res(L(z), z_0).$$

We computed the residues in two ways. One method is to use the `Residue[]` function available in Mathematica V4, which computes a series expansion of $L(z)$ symbolically. The second method is to hand-code the residue rules and use symbolic differentiation. Both methods give exact rational answers in terms of ρ_r and c_r. However, the hand-coded residue rules are significantly faster in execution time.

Bibliography

[AD76] M. B. Abrahamse and R. G. Douglas, *A class of subnormal operators related to multiply-connected domains*, Advances in Math. **19** (1976), no. 1, 106–148.

[Agl80] Jim Agler, *An invariant subspace theorem*, J. Funct. Anal. **38** (1980), no. 3, 315–323.

[Agl85] _____, *Rational dilation on an annulus*, Ann. of Math. (2) **121** (1985), no. 3, 537–563.

[Agl90] _____, *Operator theory and the Carathéodory metric*, Invent. Math. **101** (1990), no. 2, 483–500.

[And63] T. Andô, *On a pair of commutative contractions*, Acta Sci. Math. (Szeged) **24** (1963), 88–90.

[Arv69] William B. Arveson, *Subalgebras of C^*-algebras*, Acta Math. **123** (1969), 141–224.

[Arv72] William Arveson, *Subalgebras of C^*-algebras. II*, Acta Math. **128** (1972), no. 3-4, 271–308.

[AY99] J. Agler and N. J. Young, *A commutant lifting theorem for a domain in $\mathbf{C}^2$ and spectral interpolation*, J. Funct. Anal. **161** (1999), no. 2, 452–477.

[AY00] _____, *Operators having the symmetrized bidisc as a spectral set*, Proc. Edinburgh Math. Soc. (2) **43** (2000), no. 1, 195–210.

[Ber] Michel Berkelaar, *lp_solve*, Available at ftp://ftp.es.ele.tue.nl/pub/lp_solve/.

[Ber68] C. Berger, *Normal dilations*, Ph.D. thesis, Cornell University, 1968.

[BS92] M.D. Bird and C.R. Steele, *A solution procedure for Laplace's equation on multiply connected circular domains*, Transactions of the ASME: Journal of Applied Mechanics **59** (1992), no. 2, 398–404.

[Con90] John B. Conway, *A course in functional analysis*, second ed., Graduate Texts in Mathematics, vol. 96, Springer-Verlag, New York, 1990.

[Con95] _____, *Functions of one complex variable. II*, Springer-Verlag, New York, 1995.

[CW67] R. Coifman and Guido Weiss, *A kernel associated with certain multiply connected domains and its applications to factorization theorems*, Studia Math. **28** (1966/1967), 31–68.

[DM05] Michael A. Dritschel and Scott McCullough, *The failure of rational dilation on a triply connected domain*, J. Amer. Math. Soc. **18** (2005), no. 4, 873–918 (electronic).

[DP86] R. G. Douglas and V. I. Paulsen, *Completely bounded maps and hypo-Dirichlet algebras*, Acta Sci. Math. (Szeged) **50** (1986), no. 1-2, 143–157.

[DP89] Ronald G. Douglas and Vern I. Paulsen, *Hilbert modules over function algebras*, Pitman Research Notes in Mathematics Series, vol. 217, Longman Scientific & Technical, Harlow, 1989.

[Fil70] Peter A. Fillmore, *Notes on operator theory*, Van Nostrand Reinhold Mathematical Studies, No. 30, Van Nostrand Reinhold Co., New York, 1970.

[Fis83] Stephen D. Fisher, *Function theory on planar domains*, John Wiley & Sons Inc., New York, 1983, A second course in complex analysis, A Wiley-Interscience Publication.

[FK99] Stephen D. Fisher and Dmitry Khavinson, *Extreme Pick-Nevanlinna interpolants*, Canad. J. Math. **51** (1999), no. 5, 977–995.

[Foi59] C. Foias, *Certaines applications des ensembles spectraux. 1. mesure harmonique-spectrale.*, Studii Cercetari Mat. **10** (1959), 365–401.

[For79] Frank Forelli, *The extreme points of some classes of holomorphic functions*, Duke Math. J. **46** (1979), no. 4, 763–772.

[GMW81] Philip E. Gill, Walter Murray, and Margaret H. Wright, *Practical optimization*, Academic Press Inc. [Harcourt Brace Jovanovich Publishers], London, 1981.

[Gru78] Helmut Grunsky, *Lectures on theory of functions in multiply connected domains*, Vandenhoeck & Ruprecht, Göttingen, 1978, Studia Mathematica, Skript 4.

[Hei85]   Maurice Heins, *Extreme normalized analytic functions with positive real part*, Ann. Acad. Sci. Fenn. Ser. A I Math. **10** (1985), 239–245.

[Kha84]   D. Khavinson, *On removal of periods of conjugate functions in multiply connected domains*, Michigan Math. J. **31** (1984), no. 3, 371–379.

[KS78]   N. Kerzman and E. M. Stein, *The Cauchy kernel, the Szegö kernel, and the Riemann mapping function*, Math. Ann. **236** (1978), no. 1, 85–93.

[Lau73]   R. Lautzenheiser, *Spectral sets, reducing subspaces, and function algebras*, Ph.D. thesis, Indiana University, 1973.

[Leb63]   Arnold Lebow, *On von Neumann's theory of spectral sets*, J. Math. Anal. Appl. **7** (1963), 64–90.

[McC95]   Scott McCullough, *Matrix functions of positive real part on an annulus*, Houston J. Math. **21** (1995), no. 3, 489–506.

[Mis84]   Gadadhar Misra, *Curvature inequalities and extremal properties of bundle shifts*, J. Operator Theory **11** (1984), no. 2, 305–317.

[Mla72]   W. Mlak, *Partitions of spectral sets*, Ann. Polon. Math. **25** (1971/72), 273–280.

[Nai43]   M. A. Naimark, *On a representation of additive operator set functions*, C. R. (Doklady) Acad. Sci. URSS (N.S.) **41** (1943), 359–361.

[NN94]   Yurii Nesterov and Arkadii Nemirovskii, *Interior-point polynomial algorithms in convex programming*, Society for Industrial and Applied Mathematics (SIAM), Philadelphia, PA, 1994.

[Par70]   Stephen Parrott, *Unitary dilations for commuting contractions*, Pacific J. Math. **34** (1970), 481–490.

[Pau86]   Vern I. Paulsen, *Completely bounded maps and dilations*, Pitman Research Notes in Mathematics Series, vol. 146, Longman Scientific & Technical, Harlow, 1986.

[Pau87]   Vern Paulsen, *K-spectral values for some finite matrices*, J. Operator Theory **18** (1987), no. 2, 249–263.

[Pau88]   Vern I. Paulsen, *Toward a theory of K-spectral sets*, Surveys of some recent results in operator theory, Vol. I, Longman Sci. Tech., Harlow, 1988, pp. 221–240.

[Put97a] Mihai Putinar, *A dilation theory approach to cubature formulas*, Exposition. Math. **15** (1997), no. 2, 183–192.

[Put97b] _____, *Spectral sets and scalar dilations*, Houston J. Math. **23** (1997), no. 2, 247–265.

[Put00]   _____, *A dilation theory approach to cubature formulas. II*, Math. Nachr. **211** (2000), 159–175.

[Rap02]   Benjamin Raphael, *A computational investigation of spectral sets and rational dilations over multiply-connected domains*, Ph.D. thesis, University of California, San Diego, 2002.

[Rud87]   Walter Rudin, *Real and complex analysis*, third ed., McGraw-Hill Book Co., New York, 1987.

[Sar65a] Donald Sarason, *The H^p spaces of an annulus*, Mem. Amer. Math. Soc. No. **56** (1965), 78.

[Sar65b] _____, *On spectral sets having connected complement*, Acta Sci. Math. (Szeged) **26** (1965), 289–299.

[SN53]   Béla Sz.-Nagy, *Sur les contractions de l'espace de Hilbert*, Acta Sci. Math. Szeged **15** (1953), 87–92.

[Teg98]   T.J. Tegtmeyer, *The Ahlfors map and Szegö kernel in multiply connected domains*, Ph.D. thesis, Purdue University, 1998.

[Var74]   N. Th. Varopoulos, *On an inequality of von Neumann and an application of the metric theory of tensor products to operators theory*, J. Functional Analysis **16** (1974), 83–100.

[VB96]   Lieven Vandenberghe and Stephen Boyd, *Semidefinite programming*, SIAM Rev. **38** (1996), no. 1, 49–95.

[Vin98]   Victor Vinnikov, *Commuting operators and function theory on a Riemann surface*, Holomorphic spaces (Sheldon Axler, John E. McCarthy, and Donald Sarason, eds.), Cambridge University Press, Cambridge, 1998, Papers from the MSRI Program held in Berkeley, CA, 1995, pp. 445–476.

[vN51]   Johann von Neumann, *Eine Spektraltheorie für allgemeine Operatoren eines unitären Raumes*, Math. Nachr. **4** (1951), 258–281.

[Wol99]   Stephen Wolfram, *The Mathematica book*, fourth ed., Wolfram Media, Inc., Champaign, IL, 1999.

[Wri97]   Stephen J. Wright, *Primal-dual interior-point methods*, Society for Industrial and Applied Mathematics (SIAM), Philadelphia, PA, 1997.
[Zmo58]   V. A. Zmorovič, *On the generalisation of Schwarz's integral formula on n-connected circular domains*, Dopovidi Akad. Nauk Ukraïn. RSR **1958** (1958), 489–492.

Editorial Information

To be published in the *Memoirs*, a paper must be correct, new, nontrivial, and significant. Further, it must be well written and of interest to a substantial number of mathematicians. Piecemeal results, such as an inconclusive step toward an unproved major theorem or a minor variation on a known result, are in general not acceptable for publication.

Papers appearing in *Memoirs* are generally at least 80 and not more than 200 published pages in length. Papers less than 80 or more than 200 published pages require the approval of the Managing Editor of the Transactions/Memoirs Editorial Board.

As of September 30, 2007, the backlog for this journal was approximately 14 volumes. This estimate is the result of dividing the number of manuscripts for this journal in the Providence office that have not yet gone to the printer on the above date by the average number of monographs per volume over the previous twelve months, reduced by the number of volumes published in four months (the time necessary for preparing a volume for the printer). (There are 6 volumes per year, each usually containing at least 4 numbers.)

A Consent to Publish and Copyright Agreement is required before a paper will be published in the *Memoirs*. After a paper is accepted for publication, the Providence office will send a Consent to Publish and Copyright Agreement to all authors of the paper. By submitting a paper to the *Memoirs*, authors certify that the results have not been submitted to nor are they under consideration for publication by another journal, conference proceedings, or similar publication.

Information for Authors

Memoirs are printed from camera copy fully prepared by the author. This means that the finished book will look exactly like the copy submitted.

Initial submission. The AMS uses Centralized Manuscript Processing for initial submissions. Authors should submit a PDF file using the Initial Manuscript Submission form found at www.ams.org/cgi-bin/peertrack/submission.pl, or send one copy of the manuscript to the following address: Centralized Manuscript Processing, MEMOIRS OF THE AMS, 201 Charles Street, Providence, RI 02904-2294 USA. If a paper copy is being forwarded to the AMS, indicate that it is for it Memoirs and include the name of the corresponding author, contact information such as email address or mailing address, and the name of an appropriate Editor to review the paper (see the list of Editors below).

The paper must contain a *descriptive title* and an *abstract* that summarizes the article in language suitable for workers in the general field (algebra, analysis, etc.). The *descriptive title* should be short, but informative; useless or vague phrases such as "some remarks about" or "concerning" should be avoided. The *abstract* should be at least one complete sentence, and at most 300 words. Included with the footnotes to the paper should be the 2000 *Mathematics Subject Classification* representing the primary and secondary subjects of the article. The classifications are accessible from www.ams.org/msc/. The list of classifications is also available in print starting with the 1999 annual index of *Mathematical Reviews*. The Mathematics Subject Classification footnote may be followed by a list of *key words and phrases* describing the subject matter of the article and taken from it. Journal abbreviations used in bibliographies are listed in the latest *Mathematical Reviews* annual index. The series abbreviations are also accessible from www.ams.org/publications/. To help in preparing and verifying references, the AMS offers MR Lookup, a Reference Tool for Linking, at www.ams.org/mrlookup/.

Electronically prepared manuscripts. The AMS encourages electronically prepared manuscripts, with a strong preference for $\mathcal{AMS}$-LaTeX. To this end, the Society has prepared $\mathcal{AMS}$-LaTeX author packages for each AMS publication. Author packages include instructions for preparing electronic manuscripts, samples, and a style file that generates the particular design specifications of that publication series. Though $\mathcal{AMS}$-LaTeX is the highly preferred format of TeX, author packages are also available in $\mathcal{AMS}$-TeX.

Authors may retrieve an author package from the AMS website starting from www.ams.org/tex/ or via FTP to ftp.ams.org (login as anonymous, enter username as password, and type cd pub/author-info). The *AMS Author Handbook* and the *Instruction Manual* are available in PDF format following the author packages link from www.ams.org/tex/. The author package can also be obtained free of charge by sending email to tech-support@ams.org (Internet) or from the Publication Division, American Mathematical Society, 201 Charles St., Providence, RI 02904-2294, USA. When requesting an author package, please specify $\mathcal{AMS}$-LaTeX or $\mathcal{AMS}$-TeX and the publication in which your paper will appear. Please be sure to include your complete mailing address.

After acceptance. The final version of the electronic file should be sent to the Providence office (this includes any TeX source file, any graphics files, and the DVI or PostScript file) immediately after the paper has been accepted for publication.

Before sending the source file, be sure you have proofread your paper carefully. The files you send must be the EXACT files used to generate the proof copy that was accepted for publication. For all publications, authors are required to send a printed copy of their paper, which exactly matches the copy approved for publication, along with any graphics that will appear in the paper.

Accepted electronically prepared files can be submitted via the web at www.ams.org/submit-book-journal/, sent via FTP, or sent on CD-Rom or diskette to the Electronic Prepress Department, American Mathematical Society, 201 Charles Street, Providence, RI 02904-2294 USA. TeX source files, DVI files, and PostScript files can be transferred over the Internet by FTP to the Internet node ftp.ams.org (130.44.1.100). When sending a manuscript electronically via CD-Rom or diskette, please be sure to include a message identifying the paper as a Memoir.

Electronically prepared manuscripts can also be sent via email to pub-submit@ams.org (Internet). In order to send files via email, they must be encoded properly. (DVI files are binary and PostScript files tend to be very large.)

Electronic graphics. Comprehensive instructions on preparing graphics are available at www.ams.org/jourhtml/. A few of the major requirements are given here.

Submit files for graphics as EPS (Encapsulated PostScript) files. This includes graphics originated via a graphics application as well as scanned photographs or other computer-generated images. If this is not possible, TIFF files are acceptable as long as they can be opened in Adobe Photoshop or Illustrator. No matter what method was used to produce the graphic, it is necessary to provide a paper copy to the AMS.

Authors using graphics packages for the creation of electronic art should also avoid the use of any lines thinner than 0.5 points in width. Many graphics packages allow the user to specify a "hairline" for a very thin line. Hairlines often look acceptable when proofed on a typical laser printer. However, when produced on a high-resolution laser imagesetter, hairlines become nearly invisible and will be lost entirely in the final printing process.

Screens should be set to values between 15% and 85%. Screens which fall outside of this range are too light or too dark to print correctly. Variations of screens within a graphic should be no less than 10%.

Inquiries. Any inquiries concerning a paper that has been accepted for publication should be sent to memo-query@ams.org or directly to the Electronic Prepress Department, American Mathematical Society, 201 Charles St., Providence, RI 02904-2294 USA.

Editors

This journal is designed particularly for long research papers, normally at least 80 pages in length, and groups of cognate papers in pure and applied mathematics. Papers intended for publication in the *Memoirs* should be addressed to one of the following editors. The AMS uses Centralized Manuscript Processing for initial submissions to AMS journals. Authors should follow instructions listed on the Initial Submission page found at www.ams.org/memo/memosubmit.html.

Algebra to ALEXANDER KLESHCHEV, Department of Mathematics, University of Oregon, Eugene, OR 97403-1222; email: ams@noether.uoregon.edu

Algebraic geometry and its application to MINA TEICHER, Emmy Noether Research Institute for Mathematics, Bar-Ilan University, Ramat-Gan 52900, Israel; email: teicher@macs.biu.ac.il

Algebraic geometry to DAN ABRAMOVICH, Department of Mathematics, Brown University, Box 1917, Providence, RI 02912; email: amsedit@math.brown.edu

Algebraic number theory to V. KUMAR MURTY, Department of Mathematics, University of Toronto, 100 St. George Street, Toronto, ON M5S 1A1, Canada; email: murty@math.toronto.edu

Algebraic topology to ALEJANDRO ADEM, Department of Mathematics, University of British Columbia, Room 121, 1984 Mathematics Road, Vancouver, British Columbia, Canada V6T 1Z2; email: adem@math.ubc.ca

Combinatorics to JOHN R. STEMBRIDGE, Department of Mathematics, University of Michigan, Ann Arbor, Michigan 48109-1109; email: FRS@umich.edu

Complex analysis and harmonic analysis to ALEXANDER NAGEL, Department of Mathematics, University of Wisconsin, 480 Lincoln Drive, Madison, WI 53706-1313; email: nagel@math.wisc.edu

Differential geometry and global analysis to LISA C. JEFFREY, Department of Mathematics, University of Toronto, 100 St. George St., Toronto, ON Canada M5S 3G3; email: jeffrey@math.toronto.edu

Dynamical systems and ergodic theory to AMIE WILKINSON, Department of Mathematics, Northwestern University, 2033 Sheridan Road, Evanston, IL 60208-2730; email: transactions@math.northwestern.edu

Functional analysis and operator algebras to DIMITRI SHLYAKHTENKO, Department of Mathematics, University of California, Los Angeles, CA 90095; email: shlyakht@math.ucla.edu

Geometric analysis to WILLIAM P. MINICOZZI II, Department of Mathematics, Johns Hopkins University, 3400 N. Charles St., Baltimore, MD 21218; email: trans@math.jhu.edu

Geometric analysis to MLADEN BESTVINA, Department of Mathematics, University of Utah, 155 South 1400 East, JWB 233, Salt Lake City, Utah 84112-0090; email: bestvina@math.utah.edu

Harmonic analysis, representation theory, and Lie theory to ROBERT J. STANTON, Department of Mathematics, The Ohio State University, 231 West 18th Avenue, Columbus, OH 43210-1174; email: stanton@math.ohio-state.edu

Logic to STEFFEN LEMPP, Department of Mathematics, University of Wisconsin, 480 Lincoln Drive, Madison, Wisconsin 53706-1388; email: lempp@math.wisc.edu

Partial differential equations to GUSTAVO PONCE, Department of Mathematics, South Hall, Room 6607, University of California, Santa Barbara, CA 93106; email: ponce@math.ucsb.edu

Partial differential equations and dynamical systems to PETER POLACIK, School of Mathematics, University of Minnesota, Minneapolis, MN 55455; email: polacik@math.umn.edu

Probability and statistics to KRZYSZTOF BURDZY, Department of Mathematics, University of Washington, Box 354350, Seattle, Washington 98195-4350; email: burdzy@math.washington.edu

Real analysis and partial differential equations to DANIEL TATARU, Department of Mathematics, University of California, Berkeley, Berkeley, CA 94720; email: tataru@math.berkeley.edu

All other communications to the editors should be addressed to the Managing Editor, ROBERT GURALNICK, Department of Mathematics, University of Southern California, Los Angeles, CA 90089-1113; email: guralnic@math.usc.edu.

Titles in This Series

895 **Steffen Roch,** Finite sections of band-dominated operators, 2008
894 **Martin Dindoš,** Hardy spaces and potential theory on C^1 domains in Riemannian manifolds, 2008
893 **Tadeusz Iwaniec and Gaven Martin,** The Beltrami Equation, 2008
892 **Jim Agler, John Harland, and Benjamin J. Raphael,** Classical function theory, operator dilation theory, and machine computation on multiply-connected domains, 2008
891 **John H. Hubbard and Peter Papadopol,** Newton's method applied to two quadratic equations in $\mathbb{C}^2$ viewed as a global dynamical system, 2008
890 **Steven Dale Cutkosky,** Toroidalization of dominant morphisms of 3-folds, 2007
889 **Michael Sever,** Distribution solutions of nonlinear systems of conservation laws, 2007
888 **Roger Chalkley,** Basic global relative invariants for nonlinear differential equations, 2007
887 **Charlotte Wahl,** Noncommutative Maslov index and eta-forms, 2007
886 **Robert M. Guralnick and John Shareshian,** Symmetric and alternating groups as monodromy groups of Riemann surfaces I: Generic covers and covers with many branch points, 2007
885 **Jae Choon Cha,** The structure of the rational concordance group of knots, 2007
884 **Dan Haran, Moshe Jarden, and Florian Pop,** Projective group structures as absolute Galois structures with block approximation, 2007
883 **Apostolos Beligiannis and Idun Reiten,** Homological and homotopical aspects of torsion theories, 2007
882 **Lars Inge Hedberg and Yuri Netrusov,** An axiomatic approach to function spaces, spec tral synthesis and Luzin approximation, 2007
881 **Tao Mei,** Operator valued Hardy spaces, 2007
880 **Bruce C. Berndt, Geumlan Choi, Youn-Seo Choi, Heekyoung Hahn, Boon Pin Yeap, Ae Ja Yee, Hamza Yesilyurt, and Jinhee Yi,** Ramanujan's forty identities for Rogers-Ramanujan functions, 2007
879 **O. García-Prada, P. B. Gothen, and V. Muñoz,** Betti numbers of the moduli space of rank 3 parabolic Higgs bundles, 2007
878 **Alessandra Celletti and Luigi Chierchia,** KAM stability and celestial mechanics, 2007
877 **María J. Carro, José A. Raposo, and Javier Soria,** Recent developments in the theory of Lorentz spaces and weighted inequalities, 2007
876 **Gabriel Debs and Jean Saint Raymond,** Borel liftings of Borel sets: Some decidable and undecidable statements, 2007
875 **C. Krattenthaler and T. Rivoal,** Hypergéométrie et fonction zêta de Riemann, 2007
874 **Sonia Natale,** Semisolvability of semisimple Hopf algebras of low dimension, 2007
873 **A. J. Duncan,** Exponential genus problems in one-relator products of groups, 2007
872 **Anthony V. Geramita, Tadahito Harima, Juan C. Migliore, and Yong Su Shin,** The Hilbert function of a level algebra, 2007
871 **Pascal Auscher,** On necessary and sufficient conditions for L^p-estimates of Riesz transforms associated to elliptic operators on $\mathbb{R}^n$ and related estimates, 2007
870 **Takuro Mochizuki,** Asymptotic behaviour of tame harmonic bundles and an application to pure twistor D-modules, Part 2, 2007
869 **Takuro Mochizuki,** Asymptotic behaviour of tame harmonic bundles and an application to pure twistor D-modules, Part 1, 2007
868 **Gelu Popescu,** Entropy and multivariable interpolation, 2006
867 **Vilmos Totik,** Metric properties of harmonic measures, 2006
866 **William Craig,** Semigroups underlying first-order logic, 2006
865 **Nathanial P. Brown,** Invariant means and finite representation theory of $C*$-algebras, 2006

TITLES IN THIS SERIES

864 **John M. Lee,** Fredholm operators and Einstein metrics on conformally compact manifolds, 2006

863 **M. Lübke and A. Teleman,** The Universal Kobayashi-Hitchin correspondence on Hermitian manifolds, 2006

862 **Alberto Canonaco,** The Beilinson complex and canonical rings of irregular surfaces, 2006

861 **Leon A. Takhtajan and Lee-Peng Teo,** Weil-Petersson metric on the universal Teichmüller space, 2006

860 **Thomas M. Fiore,** Pseudo limits, biadjoints and pseudo algebras: Categorical foundations of conformal field theory, 2006

859 **N. Arcozzi, R. Rochberg, and E. Sawyer,** Carleson measures and interpolating sequences for Besov spaces on complex balls, 2006

858 **Enrico Valdinoci, Berardino Sciunzi, and Vasile Ovidiu Savin,** Flat level set regularity of p-Laplace phase transitions, 2006

857 **Donatella Danielli, Nocola Garofalo, and Duy-Minh Nhieu,** Non-doubling Ahlfors measures, perimeter measures, and the characterization of the trace spaces of Sobolev functions in Carnot-Carathéodory spaces, 2006

856 **Vladimir Bolotnikov and Harry Dym,** On boundary interpolation for matrix valued Schur functions, 2006

855 **Yevgenia Kashina, Yorck Sommerhäuser, and Yongchang Zhu,** On higher Frobenius-Schur indicators, 2006

854 **Noam Greenberg,** The role of true finiteness in the admissible recursively enumerable degrees, 2006

853 **Joachim Krieger,** Stability of spherically symmetric wave maps, 2006

852 **Viorel Barbu, Irena Lasiecka, and Roberto Triggiani,** Tangential boundary stabilization of Navier-Stokes equations, 2006

851 **Jie Wu,** On maps from loop suspensions to loop spaces and the shuffle relations on the Cohen groups, 2006

850 **Siegfried Echterhoff, S. Kaliszewski, John Quigg, and Iain Raeburn,** A categorical approach to imprimitivity theorems for C^*-dynamical systems, 2006

849 **Katsuhiko Kuribayashi, Mamoru Mimura, and Tetsu Nishimoto,** Twisted tensor products related to the cohomology of the classifying spaces of loop groups, 2006

848 **Bob Oliver,** Equivalences of classifying spaces completed at the prime two, 2006

847 **Eric T. Sawyer and Richard L. Wheeden,** Hölder continuity of weak solutions to subelliptic equations with rough coefficients, 2006

846 **Victor Beresnevich, Detta Dickinson, and Sanju Velani,** Measure theoretic laws for lim–sup sets, 2006

845 **Ehud Friedgut, Vojtech Rödl, Andrzej Ruciński, and Prasad V. Tetali,** A Sharp threshold for random graphs with a monochromatic triangle in every edge coloring, 2006

844 **Amadeu Delshams, Rafael de la Llave, and Tere M. Seara,** A geometric mechanism for diffusion in Hamiltonian systems overcoming the large gap problem: Heuristics and rigorous verification on a model, 2006

843 **Denis V. Osin,** Relatively hyperbolic groups: Intrinsic geometry, algebraic properties, and algorithmic problems, 2006

842 **David P. Blecher and Vrej Zarikian,** The calculus of one-sided M-ideals and multipliers in operator spaces, 2006

For a complete list of titles in this series, visit the
AMS Bookstore at **www.ams.org/bookstore/**.